Servicing Electronic Laboratory Equipment

LABORATORY MONOGRAPHS
Malignant Blood Diseases
Tissue and Cell Culture
Ultracentrifugation
Ultramicrotomy
Laboratory Planning

Servicing Electronic Laboratory Equipment

L.W. Price
M.A., C.Eng., M.I.E.R.E.

University of Cambridge

BAILLIÈRE TINDALL

© 1973 BAILLIERE TINDALL
7 & 8 Henrietta Street, London WC2 8QE
A division of Crowell Collier Macmillan Publishers Ltd.

ISBN 0 7020 0421 9

Published in the United States of America by
The Williams & Wilkins Company, Baltimore

Printed in Great Britain by
The Whitefriars Press Ltd., London and Tonbridge

Preface

The intensive use of electronic instruments in scientific laboratories is now commonplace. As a result of this, laboratory technicians are required not only to operate complex equipment but also to perform essential maintenance tasks. Many operating difficulties and instrument breakdowns are in fact often due to simple faults which can easily be rectified by a laboratory technician or scientist, with considerable savings in time and expense. In this book I have outlined the basic principles involved in 'trouble shooting' and routine maintenance on common scientific apparatus and measuring instruments.

Specialized work on instruments should be carried out only by the manufacturer's service engineer or by a laboratory technician who has attended training courses held by the manufacturer. Much damage can be caused by unqualified untrained staff attempting to rectify faults. For this reason I have clearly stated throughout the book that work which is best left to a trained electronic technician or engineer.

It is my hope that this book will be of use not only to technicians employed in hospital, research and teaching laboratories, but to scientists, research workers, and science teachers concerned with keeping their apparatus in first-class condition.

November 1972 L. W. Price

Contents

PREFACE	v
ACKNOWLEDGEMENTS	ix
CHAPTER 1: ELECTRONIC PRINCIPLES AND FAULT FINDING	1
CHAPTER 2: GENERAL LABORATORY EQUIPMENT	29
Temperature Measurement and Control	29
Electric Motors and Speed Control	41
Vacuum Systems	56
General Electronic Apparatus	58
CHAPTER 3: ELECTROCHEMICAL MEASURING AND ANALYTICAL EQUIPMENT	75
pH Electrodes and Meters	75
General Electrochemical Apparatus	86
Chromatography	93
CHAPTER 4: ELECTRO-OPTICAL MEASURING INSTRUMENTS	111
Instrument Components	111
Electro-optical Instruments	118
CHAPTER 5: RECORDING AND INDICATING DEVICES	147
Indicating Instruments	147
Recorders	150
Oscilloscopes	162
CHAPTER 6: RADIATION DETECTORS	167

CONTENTS

APPENDIX	184
Books for Further Reading	184
Data Books and Tables	185
Electrical and Electronic Data	185
Electrochemical Data	187
Spectrophotometric Data	189
Conversion Factors and Physical Data	190
INDEX	192

Acknowledgements

I would like to express my appreciation to my colleagues for helpful suggestions made in the course of the preparation of this book, to my wife Brenda for her patience and efforts in typing and correcting the manuscript, to my friends Doreen Astley and Margaret Meekins for their assistance with the Appendix and Index, and to Mr. Jewitt of the Photographic Department in the Department of Biochemistry for his assistance with photographs and diagrams.

I would also like to thank the firms listed below for their co-operation in supplying information and diagrams.

Beckman RIIC Ltd.
Britec Ltd.
Edwards Vacuum Components.
Electrolube Ltd.
Honeywell Ltd.
LKB Instruments Ltd.
Linstead Electronics.

Measuring & Scientific Equipment Ltd.
Mini Instruments.
Nuclear Enterprises Ltd.
Panax Ltd.
Perkin Elmer Ltd.
Pye Unicam Ltd.
Telequipment Ltd.

1 Electronic Principles and Fault Finding

Tools and equipment

For most servicing tasks, and for instrument calibration and testing, the basic equipment required comprises a multirange testmeter, soldering iron and a kit of tools. The testmeter should be capable of measuring both a.c. and d.c. voltages and currents and also have several ranges for resistance measurement. Two soldering irons are useful, one rated at 50 W for general work and the other at 25 W or less for work on small components and printed circuit boards. The tool kit should include a range of screwdrivers, several sizes and types of pliers, wire strippers and cutters, a set of open-ended spanners (including B.A. and metric), box spanners, allen keys and a medium-size adjustable spanner. In addition to this kit, an insulation tester such as a 'Megger' and a 'go' or 'no go' (good or bad) transistor tester are desirable.

If complex instruments are to be serviced then more extensive electronic test equipment is required. Examples of this type of equipment are an oscilloscope for monitoring waveforms, and pulse, square and sine wave generators. If an electronic servicing workshop is being established, then a transistor tester, valve tester, high-resistance electronic testmeter and a universal bridge capable of measuring resistance, capacitance and inductance would be included.

Fuses

When an item of equipment is switched on but does not work, first check that the mains supply is available. This can be verified by plugging other apparatus or a test tamp into the appropriate socket or by checking that the equipment mains-indicator lamp is on. If the supply is available at the mains socket but not apparently at the equipment, check the fuse in the mains plug and if this is satisfactory, check the mains fuse(s) in the equipment.

Fuses on electronic equipment are usually of the cartridge type, in which the fuse wire is enclosed in a glass tube with metal end caps for electrical connection. On electronic equipment the fuses protecting the circuits are of a low current rating and consequently the wire within the glass cartridge is of a small diameter. This means that the break in the thin wire of a blown fuse may be difficult to see and an electrical continuity test may be necessary to check whether in fact the fuse wire is open circuit. Normally however a blown fuse in the mains supply circuit can easily be seen and there may also be blackening inside the glass cartridge.

On occasions, a fuse wire may break due to fatigue and this can be caused by the surge current that flows when switching on. In such situations it is only necessary to replace the fuse. A fuse blown by a short-circuit fault in the equipment, usually blackens the inside of the glass and the wire material may also be sputtered on the inner glass surface. In such cases always look for the fault before replacing the fuse. It should be noted that on some equipment where high surge currents flow at switch on, special 'slow blow' or anti-surge fuses are used. They allow high currents to flow for a very short period without rupturing the fuse.

Should only part of an item of equipment work, or its operation be obviously incorrect, then fuses other than those in the mains power circuit may have blown. Such fuses are often situated at the rear of electronic instruments and are sometimes internal. It is important only to replace a fuse with one of the same type and rating. Do not substitute a slow blow fuse for the same current rating normal type.

Physical indications of faulty components
Should a replacement fuse blow, the next step in a logical fault finding sequence is to switch off and unplug from the mains socket. Next remove the screws holding the front panel of the equipment to the case or cabinet. The electronic chassis and front panel assembly can then be removed for visual examination.

A burnt or hot smell usually indicates a short circuit on the equipment and may emanate from a transformer which would be hot to touch. The protective varnish, wax or bitumen coating over the transformer winding may also have melted. The fault can be due to shorted turns in the transformer winding but the possibility of a short circuit in the components supplied from the transformer should not be overlooked.

A heavily overloaded resistor will become blackened and burnt and its surface crazed. The resistor must of course be replaced but the fault which caused the overload must first be found. This might for instance, be due to a faulty transistor or capacitor. A faulty resistor must be replaced with one of the same ohmic value which is usually indicated by coloured bands round the body of the resistor. The resistance colour code is illustrated in Figure 1.1 though precision resistors often have the resistance value hand written on the body. The physical size of a resistor depends on its wattage rating, and if a resistor is being replaced, ensure that the replacement is of the same (or higher) wattage rating.

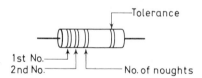

Figure 1.1. Resistance colour code.

Resistance colour code

Colour	Number	Colour	Number
Black	0	Green	5
Brown	1	Blue	6
Red	2	Violet	7
Orange	3	Grey	8
Yellow	4	White	9

Tolerance Bands (the absence of a band indicates a 20 per cent tolerance)

Gold	5%
Silver	10%
Pink	1%

Example: Coloured bands of green, blue and yellow, read from left to right indicate a 564 kΩ resistance.

A faulty capacitor may show signs of melted wax, corrosion, leakage, or have a bulging cyclindrical metal case, depending on its type and construction. When replacing a suspect capacitor of a particular capacitance value (usually expressed in pico- or microfarads), check that the working voltage of the replacement is the same as, or higher than, that of the suspect component. Most capacitors have the capacitance value and voltage rating written or printed on them. Some manufacturers however use a colour code and examples are shown in Figure 1.2.

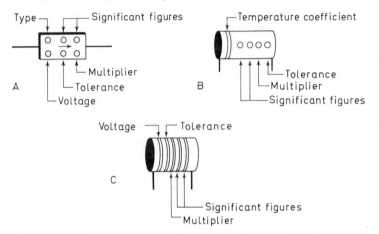

Figure 1.2. Capacitor colour codes

When replacing electrolytic capacitors it is very important to ensure that the polarity of the applied voltage is correct. If it is not, an internal leakage current flows giving rise to heat and high internal pressure resulting in an explosion. This polarized type of capacitor can only be used on direct current circuits and the positive terminal must be connected to the positive side of the supply. Damage can also result if the maximum voltage rating is exceeded or if an alternating voltage is applied. Capacitors of this type are often used for smoothing the output from rectifier circuits (which convert an alternating voltage to a direct voltage). In this application, it is important that the value of the 'ripple' on the d.c. output voltage level does not exceed the ripple current rating of the capacitor.

Power-supply capacitors may hold a considerable charge after the mains supply is switched off and it is wise, in order to avoid the possibility of an electric shock, to discharge the capacitor terminals through a low resistance before working on the equipment.

Scorched and burnt wires or components may also be in evidence and may be due to a breakdown in the electrical insulation. It should also be remembered that damage may have been caused by a fault on an item external to the equipment. This is illustrated in Figure 1.3 which shows the aftermath of a short circuit in the load supplied from a transistor amplifier. In Figure 1.4 the effect of insulation breakdown due to moisture on a plug and socket is shown; this has resulted in arcing and a considerable portion of the component has been burnt away.

Valve operation and fault symptoms

In a valve, a heated cathode emits negatively charged electrons which are attracted to a positively charged anode plate. These electrodes are enclosed in an evacuated glass envelope as shown in Figure 1.5(A). If the anode is made negative with respect to the cathode then the electrons are repelled away from the anode and there is no current flow through the device. This type of valve is known as a diode (two electrode) and its one way action enables it to be used as a rectifier to convert an alternating current to a direct or undirectional current since current only flows when the anode is positive.

A control grid (wire mesh), through which electrons can pass, may be positioned between the cathode and anode and enables the anode current of a valve to be controlled. This is shown in Figure 1.5(B). A small negative potential on the grid repels some of the electrons (remember like charges repel, unlike attract) and the net anode current is reduced. A more positive grid potential assists the positive anode and the current increases. This type of valve is called a triode (three electrode). Extra grids were introduced to give improved performance and characteristics, resulting in the pentode valve (five electrode). The pentode valve anode current-anode voltage characteristic curves for various values of grid voltage are shown in Figure 1.5(C). In signal

SERVICING ELECTRONIC LABORATORY EQUIPMENT

Figure 1.3. Damage due to a short circuit on the output of an amplifier.

Figure 1.4. Plug and socket damaged by arcing.

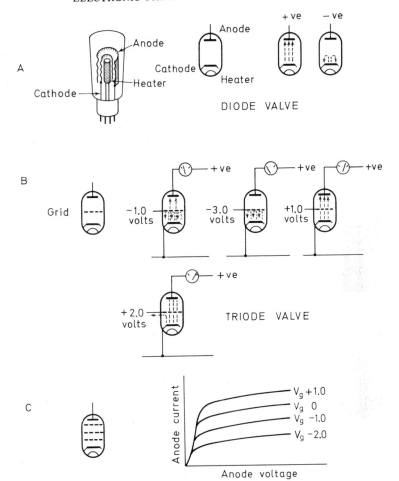

Figure 1.5. Valve operation and characteristics.

amplifying circuits the valve is biased so that the working point is in the centre of the characteristic curves. The bias voltage is obtained by utilizing the voltage drop developed across a resistance inserted in the cathode circuit. In a.c. amplifiers, a bypass capacitor is connected across the cathode resistance. The capacitor offers very low opposition to the flow of the alternating current component of the valve current and bypasses it away from the bias resistance. This is done to prevent degeneration or loss of gain.

A small change in the signal voltage applied to a valve grid causes a change in the anode current which, in flowing through a load resistance connected in the anode circuit, results in a much larger but corresponding voltage change at the anode. The variations in anode current due to the signal applied to the grid are superimposed on the steady quiescent or standing current due to the bias.

In fault finding procedures the quiescent anode current of a valve can be found by measuring the voltage drop produced across the anode load and dividing this measured value by the ohmic resistance of the load.

The cathode heater normally glows a dull red and will warm the glass envelope of the valve. A cold valve with no heater glow may thus have an open circuit heater. Power rectifier valves (nowadays found mostly on old equipment) may be filled with gas and an erratic flashing light from the valve signifies a fault.

Other types of valve or electron tube do not have a heated cathode and are known as cold cathode valves. They are typically filled with neon gas and are used as voltage stabilizers. Stabilizers usually give a steady red or blue glow (depending on the gas filling) and flashing indicates a faulty valve. This can be verified by connecting a testmeter, switched to measure d.c. voltage, across the anode and cathode base connections. If erratic readings are obtained, the stabilizer should be replaced.

On valves, the various electrodes are nearly always connected to pins on the base of the glass envelope. The valve can then be plugged into a socket mounted on the electronic chassis. Numbers on the circuit diagram symbol adjacent to the valve electrodes, indicate the pins to which they are connected. A diagram may also be given of the valve base. If the instrument manufacturers diagrams do not give this information, but only the valve type number, the connections may be found by referring to valve manufacturers data or a handbook such as the *Radio Valve Data* book published by Iliffe.

Typical faults encountered on valves are open circuit heaters, a short circuit between heater and cathode, low insulation between electrodes, low thermionic emission from the cathode and faulty sealing between the glass envelope and valve base allowing air to enter and the valve to become soft.

Semiconductor principles and practice

Semiconductor devices such as diodes, transistors and thyristors are composed of N-type and P-type semiconductor crystal lattice materials (e.g. germanium and silicon) arranged in various ways. The N-type material has an excess of electrons and therefore a net negative charge. With the P-type material there is a deficiency of electrons in the atomic structure resulting in 'positive holes', the material having a net positive charge. A PN junction diode allows current to flow easily in one direction but offers high resistance to current flow in the opposite direction. This is shown in Figure 1.6(A). Free electrons are in the majority in N-type material although there are also some positive holes present. Similarly, in P-type material, positive holes are in the majority with some minority electrons. In the forward bias condition the majority current carriers are repelled away from the ends of the material and cross the junction to give a high current flow. With reverse bias, only the minority carriers (small in number) are repelled away from the ends to cross the junction giving a small current. The characteristics of the diode are shown in Figure 1.6(A). Another diode with which we shall be concerned is the zener diode. This type of semiconductor diode operates in the region of the reverse voltage characteristic where an almost constant voltage is obtained for different current values. They are available in standard voltage ratings.

Conventional transistors are composed of semiconductor material arranged in NPN or PNP forms, the centre part being of thin cross-section. In practice, the semiconductor elements are mounted on a supporting structure and enclosed in a metal can or encapsulated in an epoxy resin moulding. Three leads are normally brought out to enable electrical connections to be made. Referring to Figure 1.6(B), which shows the operation of a PNP transistor, we see that with the potentials applied as shown, positive holes

SERVICING ELECTRONIC LABORATORY EQUIPMENT

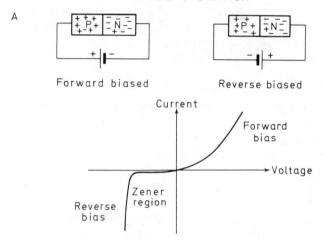

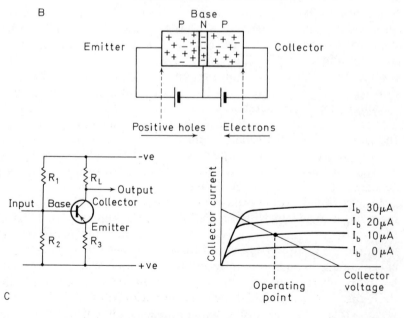

Figure 1.6. Semiconductor operation.

travel from the emitter section towards the centre base region. Rather than take the high resistance return path down the base section, they travel across the thin-width base region to the collector. A similar argument applies to the electrons moving away from the collector end. How many of these majority current carriers overcome the potential barrier of the base region depends on the base bias potential applied. By changing the emitter–base potential we can control the emitter–collector current. There is of course current flowing in the base region but about 98 per cent of the current from the emitter reaches the collector.

In servicing work it is important to remember that with the normal transistor amplifier, the emitter–base junction is always forward biased and of low resistance, while the base–collector is reverse biased and of high resistance. This is true for both PNP and NPN transistors. A typical basic amplifier circuit, known as the common emitter connection is shown in Figure 1.6(C). The load resistance R_L in the collector circuit can be of a high value (say 10kΩ) and the change in collector current I_c caused by a change in base current I_b gives a high current gain. The current gain β for this common emitter amplifier is given by $\Delta I_c / \Delta I_b$.

The operating point on the characteristic curve is determined by bias resistances R_1 and R_2. Resistance R_3 in the emitter circuit helps to stabilize the operating point against changes in temperature; its value is usually low.

Transistors are very sensitive to temperature and care is necessary when soldering or unsoldering the connecting leads. Long-nosed pliers should be used to grip the wire lead between the transistor and soldering point as shown in Figure 1.7. The pliers act as a heat shunt and prevent heat being conducted to the transistor. Clip-on heat shunts can also be obtained.

Not all transistors have three connecting leads as some types (typically power transistors) have only two, the metal case acting as the third connection. Semiconductor power output devices are normally mounted on a heat-dissipating surface known as a heat sink, which is usually made of aluminium and may have fins to dissipate as much heat as possible. Insulating bushes prevent the power transistor leads from touching the metal heat sink and a mica washer insulates the metal body or case. This insulation is important or a short circuit to the chassis will result.

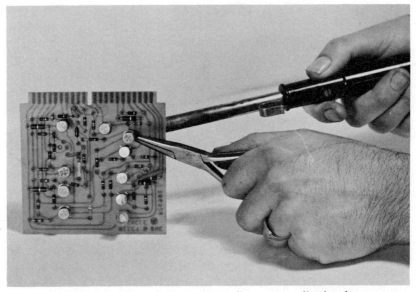

Figure 1.7. Use of heat shunt (long-nose pliers) when unsoldering a transistor from a printed circuit board.

When searching for physical signs of a fault in equipment, the temperature of a semiconductor can or case can give an indication of where the trouble is located. A cold power transistor can indicate an open circuit; a very hot normal transistor may indicate an internal fault or heavy overload current passing through it.

It should be noted that transistors are manufactured in bulk and considerable variations in parameters such as gain and leakage current may occur. For this reason it is not possible to list exact equivalent types although, depending on the circuit in which they are used, transistors with similar characteristics can sometimes be substituted. In some instrument circuits, transistors may be selected for a specified parameter, and in the event of a fault, the substitution of an unselected transistor of the same type number may affect the instrument's performance.

There are many types of semiconductor device and it is impossible in a book of this nature to describe them all. Common types and the circuits in which they are used will however be briefly described in the following chapters.

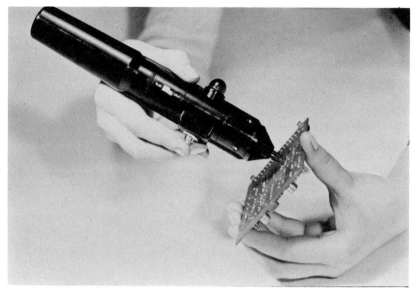

Figure 1.8. Using a transistor tester (courtesy of Britec Ltd.).

Testing transistors

Transistors can either be tested *in situ* or preferably disconnected from the circuit and tested separately. Before testing *in situ* or disconnecting a transistor from a circuit ensure that the power supply is switched off. Simple transistor testers are of two types. The first uses an indicator lamp to show whether a diode or transistor is open or short circuit. An example of this type is shown in Figure 1.8. Three spring-loaded pins are used for *in situ* measurements while for unmounted transistors an extension socket with flying leads is connected. With the tester connected, a short circuit between emitter and collector will cause the indicator lamp to light. If this test is satisfactory and the lamp does not light, then the probe push button is pressed. The lamp should now light indicating that the transistor is conducting and good. Failure of the lamp to light when the button is pressed indicates an open circuit transistor. When this type of tester is used for *in situ* measurements it is important that the external resistance of the base-emitter circuit exceeds 1kΩ for silicon transistors and 600Ω for germanium transistors. If it does not, then the tester will indicate a short circuit and the indicator lamp will light when the probe is applied.

The second type of tester uses a calibrated meter for indicating the transistor leakage current and current gain. The leakage current (I_{co}) should be less than a certain value (specified in μA in the manufacturers data) and is a measurement of the collector current with the collector junction reverse biased and the emitter open circuit. The direct current gain is usually measured by switching a resistance into the transistor base circuit to give a known fixed base current, and measuring the collector current on the meter scale calibrated in terms of gain. When connecting a transistor to a meter tester it is important that the connections are correct for the type of transistor being tested. The collector lead is connected to the negative terminal for a PNP type and to a positive terminal for an NPN type.

Transistors are not only damaged by the application of excess heat but by excess voltage and by connecting voltages of the wrong polarity. If when connected into circuit, a PNP transistor has the positive side of the voltage supply connected to the collector, the collector base junction is forward biased and thus of low resistance. A high current then flows which may produce sufficient heat to disrupt the semiconductor crystal lattice structure and destroy the transistor. Surge currents can also damage transistors and it is for this reason that the power supply must be switched off before disconnecting or connecting a transistor into circuit.

Circuit diagrams and colour codes

Before test equipment is used in tracing a fault, it is important that the symbols on the circuit diagram are understood. Scientific laboratory equipment may originate from many countries and since not all countries use the same symbols, the illustrations in Figure 1.9 are not confined to British Standard practice but include foreign symbols. It should also be noted that when a new semiconductor device, for instance, is introduced, there may be no

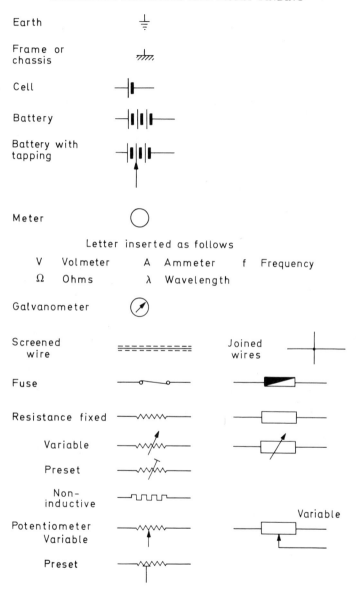

Figure 1.9. Circuit diagram symbols.

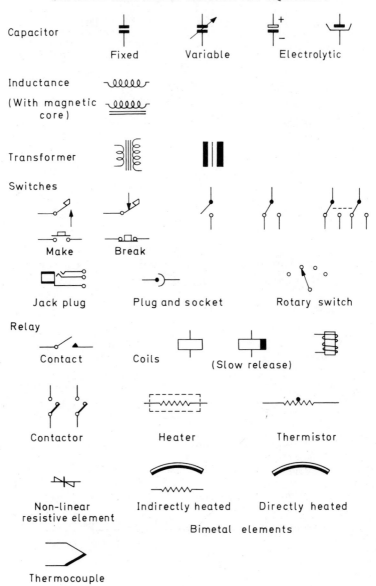

Fig. 1.9.

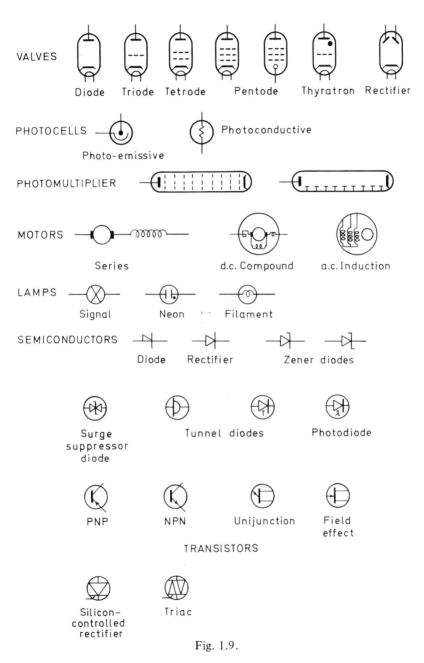

Fig. 1.9.

agreed standard symbol and a variety of the manufacturers own symbols may exist.

While the new European standard colour code for mains cables gives brown for the live lead, blue for the neutral and green/yellow for earth, much equipment in use will use old colour codes. On British equipment, red was live, black the neutral and the earth green. Some continental countries however used red for earth and when connecting up equipment of foreign manufacture, always check which colour wire is connected to the frame or earth of the instrument. American manufactured equipment uses white for live, black for neutral and green for earth.

Using a multirange testmeter

In this type of testmeter, various measuring functions and ranges are selected by means of rotary switches. The basic circuits involved are shown in a simplified form in Figure 1.10. To use an

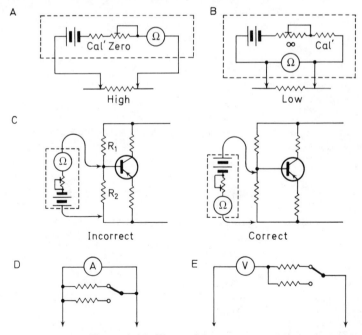

Figure 1.10. The multirange testmeter.

ohmmeter and measure resistance, a known voltage is applied to the unkown resistance, and the current flowing through the meter then gives a measure of the unknown's ohmic value. This follows from Ohm's law ($V = I.R$; where V = voltage, I = current, and R the resistance in ohms). To measure medium and high resistance values the circuit of Figure 1.10(A) is usual. The resistance range is determined by the calibration resistance value and the meter is set to indicate zero ohms (full scale deflection on meter) by adjusting the fine zero potentiometer with the test leads shorted together. To measure low resistance values the circuit of Figure 1.10(B) is used. Full scale deflection on the meter is obtained when the test leads are separated and is accurately set by the infinity adjustment potentiometer. When the test leads are shorted the meter current is zero.

Care must be exercised when measuring resistance values on transistor circuits as a voltage is applied to the circuit by the testmeter. In Figure 1.10(C), the base bias resistance R_2 is being measured, and with the testmeter negative lead on the base of the PNP transistor the base-emitter junction is forward biased and the meter indicates almost zero resistance. The correct reading is obtained if the testmeter leads are reversed. Note: On some testmeters, the polarity of the small potential applied to the circuit from the meter when measuring resistance, is opposite to that marked on the meter terminals. The terminal markings indicate the polarity of voltage or current applied to the meter for measurement.

To measure direct currents, several measuring ranges are obtained by switching various shunt (parallel) resistances across the meter as shown in Figure 1.10(D). If the meter current for full scale deflection is 1 mA, then in order to measure 1 A it is necessary to divert 999 mA through the low resistance shunt. For several ranges of voltage measurement, appropriate multiplier (series) resistances are switched into circuit as shown in Figure 1.10(E). If the 1 mA full scale deflection meter has a resistance of 100 Ω, then with no multiplier resistance in circuit, full scale deflection is obtained with a voltage of 100 mV applied. To measure 1 V, the total circuit resistance must be 1000 Ω and the multiplier resistance is thus 900 Ω.

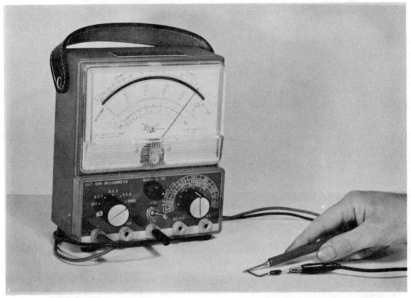

Figure 1.11. Multirange testmeter used to check resistance of a diode.

If the total internal circuit resistance of a testmeter on voltage ranges is low (say 1000 Ω/V) then the true measurement of voltage in a valve or transistor circuit will be affected. This is because the testmeter loads the circuit and causes a drop in voltage. Even with much higher resistance meters (20 000 Ω/V) a small error exists. In Figure 1.11 a multirange testmeter is being used to measure the reverse resistance of a diode. A normal diode will show a low resistance in the forward bias condition and a high resistance in the reverse bias condition.

Using an oscilloscope

An oscilloscope (the principle of which is outlined in Chapter 5) can be used to measure alternating or pulse waveforms at various points in a circuit. A common method of fault location is the use of an oscilloscope for signal tracing. For example, if in an amplifier circuit, loss or distortion of the output voltage waveform occurs, the oscilloscope is used to monitor the path, through the circuit, of a test signal applied to the amplifier input. The point in the circuit at which the signal is lost or appears distorted is then quickly traced.

Changes in the level of d.c. voltages can also be measured with an oscilloscope and a switch is provided to change from a.c. to d.c. input coupling.

All oscilloscopes have switched sensitivity ranges for the vertical deflection Y axis amplifiers and a range of X axis time base speeds to accommodate different frequencies of alternating signals. When taking measurements of waveform amplitude, the selected sensitivity setting in volts per division is multiplied by the number of vertical divisions occupied by the waveform on the screen graticule. In measuring frequency, the time base switch setting in time per division is multiplied by the number of horizontal graticule divisions for one complete cycle of the waveform. The frequency in cycles per second is then calculated from $1/T$ where T is the time for one cycle in seconds. It should be noted that incorrect results will be obtained when measuring if variable controls associated with the switch selected time base or Y amplifier ranges are not fully at one end of their travel, in the position usually marked CAL (for calibration). The fact that a testmeter may alter conditions in a circuit has already been referred to and the same applies to an oscilloscope. An oscilloscope has an input impedance typically represented by a resistance of 1 MΩ in parallel with a capacitance of about 50pF. Thus connecting an oscilloscope can alter circuit conditions giving rise to measuring errors, for example on low-current circuits and high-gain transistor circuits using megohm resistances.

A probe unit may be used with the oscilloscope leads in order to obtain a higher input impedance (e.g. 10 MΩ). It must be set up with the oscilloscope Y amplifier that is to be used for measurements. To set up the probe, a calibration (test) square-wave output from the oscilloscope (or signal generator) is applied through the probe to the Y amplifier input and the adjustment on the probe set to give an undistorted waveform on the screen. The probe can then be used for circuit measurements. When using a double-beam oscilloscope in the external trigger mode, a signal, which may or may not be one of those being monitored, is

SERVICING ELECTRONIC LABORATORY EQUIPMENT

connected to the trigger circuit input. It is important to ensure that the oscilloscope trigger circuits do not load the circuit being tested and cause a deterioration in performance.

Long oscilloscope leads may pick up hum voltages which can be coupled into the circuit being tested. The capacitance of long test leads may cause oscillation in the circuit and this can usually be prevented by connecting a 4.7 kΩ resistance in the circuit end of the non-earthing oscilloscope test lead. There is then, however, the danger that high frequency oscillation (not due to the test leads), in an amplifier for instance, may not show on the oscilloscope.

Logical fault finding

Before attempting to repair a fault on equipment make sure that its operation is understood and that you have read the manufacturers instruction manual. The manual usually contains circuit diagrams and possibly fault-finding charts or guides. The circuit diagram may also give information on the voltages and waveforms to be expected in normal operation.

Fault finding guides are of several types and possible faults may be arranged in a descending tree pattern or in a tabular form. Another more recent idea is to use a trouble-shooting flow diagram. In this method the procedure to be followed, after particular circuit checks indicate a good or bad condition, is illustrated on the diagram. Honeywell use this type of diagram in the manual on their laboratory recorder.

Information presentation techniques may use a functionally identified maintenance system and a symbolic integrated maintenance manual. These techniques use blocked schematic diagrams facing opposite to a blocked text diagram. A maintenance dependency chart, which shows the relationships between the functional items comprising the equipment, is useful for fault finding.

The exact fault-finding procedure to be followed depends on what the equipment is and on the nature of the fault. Some general points can however be made.

(1) Check fuses.
(2) If a fuse has blown, replace it with one of the same rating.

(3) If the replacement fuse blows, switch off and examine the electronic chassis for physical signs of a fault (e.g. burnt resistors, leakage from capacitors, overheated insulation and damaged insulation on wiring).

(4) When equipment fails to operate or functions erratically, and the fuses are intact, systematically check the circuit using a testmeter or oscilloscope. Generally, power-supply voltages are checked first followed by measurements on the circuit, working from the input to the output.

NOTE: Care is of course necessary, in order to avoid an electric shock, when working on live equipment removed from its case. Do not attempt what you do not understand.

The practice of mounting different sections of an electronic circuit on plug-in printed circuit boards has led to new fault finding procedures. Instead of trying to locate and repair a faulty component, logical reasoning traces the fault to a particular board. The whole board is then replaced and the faulty board returned to the instrument manufacturer or repaired at leisure. With the advent of integrated circuits, in which the equivalent of many resistors and transistors are contained in one small rectangular plug-in module, a fault necessitates replacement of the module. Integrated circuits cannot be repaired and many such units can be accommodated on one printed circuit board. There are also essential circuit components such as amplifier feedback resistances, offset current potentiometers, and frequency compensation resistances and capacitors mounted on the board. It is important to check that these items are not the cause of a malfunction before replacing an integrated circuit. In order to avoid damage due to using a non-isolated soldering iron, switch off the power supply before soldering on an integrated circuit.

A fault on electronic equipment sometimes develops only after the circuit components have warmed up and aerosol freezers or cold sprays are useful in this instance. Resistances and capacitors for example may go open or short circuit due to heating effects and the rapid cooling produced locally when the component is sprayed causes the fault symptoms to disappear. Quick tracing of faults is possible this way. A fault due to a dry soldered joint can be located by this method as the solder when cold contracts and grips the component leads, temporarily restoring continuity. The degree of local cooling is controlled by the distance of the spray nozzle from the component and the duration of the spray.

Probably the most difficult faults to trace are those occurring intermittently. These are often due to badly soldered joints, hairline cracks on printed circuit boards, and intermittent electrical connection between one or more contacting surfaces of a plug and socket. Intermittent and erratic operation can also result from dirty switch or relay contacts. All of these causes however, give rise to an intermittent variable high resistance in the circuit. Intermittent blowing of fuses sometimes occurs when chaffed or damaged wiring insulation allows a live conductor occasionally to touch the earthed frame or case of the equipment.

Electrical contacts

Switching relays and contactors are all devices which make and break electrical circuits by the opening or closing of special contact surfaces. These contacts inevitably become worn with use, particularly if heavy currents or inductive loads are switched, and burnt and pitted contacts are a source of equipment malfunction. Electrical contacts which are not too badly worn can be cleaned using a fine contact file, although very badly burnt contacts necessitate replacement of the switching device. In use, sparking can be inhibited by the use of contact lubricants. Normally when two contacts separate, the air film between them breaks down and a momentary high-temperature arc is formed. With a contact lubricant such as Electrolube, the film between the contacts is drawn out to form a bridging column between the contacts. When this breaks, the contact current is small since the bridge acts as a steadily increasing resistance. Consequently there is little or no arcing. The effect of arcing on dry contacts is shown in Figure 1.12(A) and the beneficial effects of contact lubrication in Figure 1.12(B). Contact lubricant has a low resistivity and thus reduces the resistance between the closed contact surfaces. It should be applied sparingly as excess liquid if allowed to flow over switch insulation (e.g. on a wafer switch) will give a low-resistance leakage path.

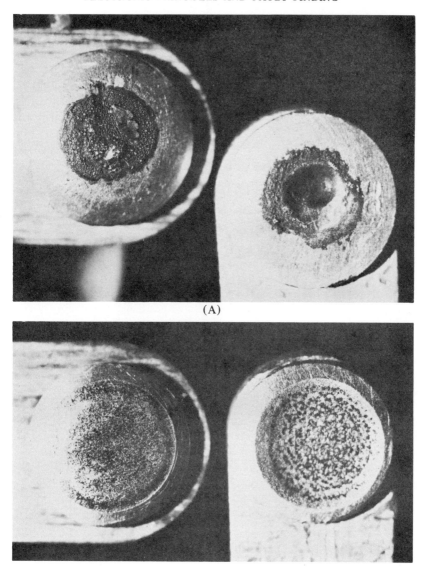

Figure 1.12. Electrical contacts (courtesy of Electrolube Ltd.): (A) wear on untreated contacts; (B) effect of contact lubricant.

Table 1.1. Properties of Electrolube contact cleaners and lubricants (courtesy of Electrolube Ltd.)

No.	Type	Applications	Temperature range	Notes
2AX	Aerosol	Timer mechanisms, battery connections, control wire pulleys, carbon commutator brushes, co-axial plugs and sockets, small gears, potentiometer tracks, thermostat contacts	−20 to +200°C	Safe on common plastics (e.g. P.V.C., polystyrene), rubber, paint
2X	Liquid			
No. 2	Liquid	Superseded by 2AX and 2X except where viscosity and wider temperature range are important	−46 to +240°C	Softens plastics
No. 1	Liquid	Useful for cleaning and lubricating non-sparking contacts. Can be used to remove hardened oil or grease films prior to applying recommended lubricant	−46 to +240°C	Softens plastics. With sparking contacts a carbon deposit is formed on contacts unless solvent is allowed to evaporate before passing current
2GX	Grease	Large electrical contacts with wiping action, knife switches, slow motion gears, bearings, push–pull rod mechanisms, large gears, heat sinks (thermocoupling medium), ultrasonics (coupling medium), oven heater element connections, cams, prevention of electrolytic action	−40 to +200°C	Safe on common plastics
2GAX	Aerosol			

ELECTRONIC PRINCIPLES AND FAULT FINDING

Some switch and contact cleaners should be used with caution as the solvents in them dissolve plastic materials. The application of various contact cleaning and lubricating liquids and greases is given in Table 1.1.

If it is necessary to replace a switch or relay, ensure that the replacement has a current rating suitable for the circuit concerned.

Insulation testing

Continuity tests can be satisfactorily carried out using an ohmmeter or testmeter switched to measure resistance. The soundness of the insulation of wiring or components to chassis or earth cannot however be tested by this means as there is only a low voltage battery within the testmeter. The testmeter would show an absolute short circuit but wiring carrying mains or higher voltages should be tested with an insulation tester such as a 'Megger'. The Megger contains a small handle-driven generator producing 250 V or 500 V and a meter calibrated in megohms. The handle is normally rotated at about 160 rev./min to produce the required test voltage. The small leakage current flowing through the insulation under test causes a deflection on the megohm calibrated meter. Modern insulation testmeters may utilize a battery-powered transistor-operated voltage converter to produce the required 500 V in place of a handle-driven generator. The test voltage is applied by pressing a button. The use of an insulation tester on apparatus is shown in Figure 1.13. Failure of electrical insulation often results in breakdown to earth and in a properly protected circuit the fuses would blow. Equipment in hazardous situations or exposed to extreme environmental conditions should be regularly checked and a record kept of the insulation resistance values. A deterioration in the insulation can then be investigated and failure due to insulation breakdown prevented. Special insulation resistance test sets are available for use on high-voltage apparatus and can apply several thousand volts. The wiring in electrophoresis installations is a laboratory application for this type of test equipment.

Routine maintenance

Regular routine maintenance is of the utmost importance if an instrument is to be maintained in first class condition. This may be

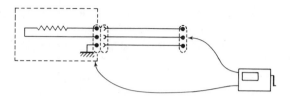

Testing insulation to earth

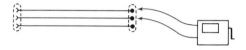

Testing insulation between leads

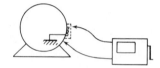

Insulation testing a motor

Figure 1.13. Insulation testing.

carried out at intervals of say three or six months and would include cleaning, lubrication of moving parts, calibration checks, internal adjustments to ensure operation is within specification limits, replacement of motor brushes, replacement of internal batteries and other components which may have deteriorated.

2 General Laboratory Equipment

TEMPERATURE MEASUREMENT AND CONTROL

Thermostat control circuits

The simplest temperature regulating device is the bimetal thermostat shown in Figure 2.1(A). The two dissimilar metals in the bimetal strip expand at different rates when subjected to heat, and the curvature of the strip then changes so as to break the electrical circuit at a preset temperature. The well known simmerstat, illustrated in Figure 2.1(B), also uses a bimetal strip but works on a different principle from a normal thermostat. A built-in miniature heating element heats the bimetal strip so that a pair of contacts are made to open and close periodically. The contacts control the supply of power to the heating element being controlled. The miniature heater is also in series with the electrical contacts and the length of time the contacts remain open or closed is determined by the setting of a control knob. The power to the load is thus controlled on a time basis.

In use, apart from deterioration of the electrical contacts, incorrect operation may result if the internal heat-adjusting screw and contact gap adjustment are not set correctly. The 'spring' of the metal strip to which the moving electrical contact is attached can change with time and the contacts may not then make and break correctly.

Another common thermostat is the stem type where a longitudinal movement, actuating the switch contacts, is obtained by the difference in expansion between the stem tube (normally brass or bronze) and an invar rod inside it. Its principle is shown in Figure 2.1(C).

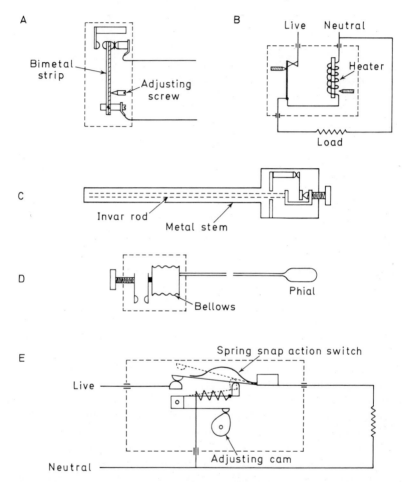

Figure 2.1. Temperature-regulating devices.

Refrigeration systems often use a bellows type thermostat. In this thermostat, shown in Figure 2.1(D), a volatile liquid (e.g. ethyl chloride) contained in a phial or tube is connected to a small metal bellows by capillary tubing. Movement of the bellows operates the switch contacts. Damage or puncturing of the bellows or capillary tubing will of course result in loss of vapour and stop the thermostat working. In laboratories, corrosion of the capillary by chemicals may be a problem.

Heating elements do occasionally fail, but in practice most trouble on heating circuits is caused by the thermostat. Older type thermostats in which the current is switched by open electrical contacts are subject to failure by the contacts becoming burnt and corroded. Modern thermostats often use a microswitch actuated by the thermostatic element. The microswitch usually has change-over electrical contacts totally enclosed within the body of the switch. They require very little mechanical effort to operate them and are small in size. The whole switch must of course be changed in the event of a failure.

Energy regulators

Electro-mechanical energy regulators which are basically the same as simmerstats described above, use a bimetal strip on which is wound a small heating element. The ratio of ON time to OFF time is set by an adjustable cam. The circuit is broken by the contacts opening due to the bending action of the bimetal strip. The principle is shown in Figure 2.1(E). Failure to regulate the power to the load may be due to the bimetal heater going open circuit. In this instance the load is permanently connected to the supply as the contacts cannot then open.

Hot-wire vacuum relays

These devices comprise an evacuated glass tube containing heavy contacts capable of switching a large current (30 amps typically). The contacts are actuated by the expansion of a wire through which a small control current (typically 25 mA) passes. Hot wire vacuum relays are very reliable but excessive current allowed to flow through the control circuit will burn out the actuating wire.

Contact thermometer circuits

In scientific laboratories, accurate temperature control of water-baths is often required and the normal type of thermostats described above are not sufficiently accurate. Instead a contact thermometer is used. Mercury rising with temperature inside the thermometer tube makes electrical contact with a fine wire, the position of which is adjustable. Another electrical connection is made to the mercury at the bottom of the thermometer, so that at a preset known temperature, the mercury acts as a switch. Very little current can be switched by the mercury contact thermometer so an amplifying system is used to control the heater current. A solid state circuit is shown in Figure 2.2.

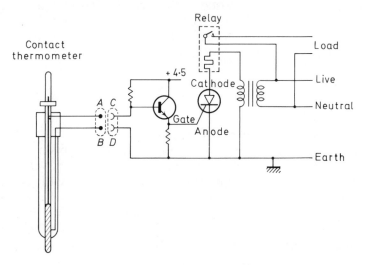

Figure 2.2. Solid-state contact-thermometer circuit.

This circuit introduces a semiconductor device known as a silicon-controlled rectifier (S.C.R.) or thyristor. This device can control a large amount of current or power by means of a small signal change at its input 'gate' connection. With the contact thermometer circuit open (temperature not at the required value) the transistor is forward biased and therefore conducts. This causes a positive signal to be applied to the gate of the S.C.R. which is turned on (i.e. made conducting). Power is then applied to the heater and the temperature of the water-bath rises. At the required temperature the circuit of the contact thermometer is made, and the transistor base-emitter circuit becomes reverse biased so that the transistor is cut off (i.e. non-conducting). The S.C.R. gate is then made negative and the device is turned off when the alternating anode voltage passes through zero. Power is then removed from the heater. It should be noted that an S.C.R. with d.c. applied to the anode-cathode circuit will remain conducting once turned on or triggered, even though the initiating gate signal is removed. In this case the anode circuit must be opened to turn off the S.C.R. In our circuit, a.c. is applied to the anode-cathode circuit and hence the device turns off when the positive half cycle of the anode waveform becomes zero. The silicon-controlled rectifier is the semiconductor equivalent of the thyratron or gas filled triode valve which it has now largely replaced.

If the control circuit does not function correctly, resulting in the water-bath overheating, check that the contact-thermometer circuit is continuous. The circuit is checked by connecting a testmeter, switched for resistance measurement, across the points A and B shown in Figure 2.2. If an open circuit is indicated (infinite resistance), the operation of the thermometer should be checked and the cable connection between the thermometer and the input plug examined for broken or disconnected wires. If the resistance check above indicates a continuous circuit (very low resistance) then short circuit C and D on the input socket and check, with the testmeter switched to measure volts, for proper operation of the transistor. The emitter voltage will be almost at the supply potential (+4.5 V) if the transistor is conducting. If the transistor is satisfactorily cut off, check whether the relay is operating by temporarily removing and then replacing the short circuit across the input socket. If the relay does not operate, the fault may be in the relay unit itself or in the S.C.R. If the relay operates, but the power circuit does not open, then check that the relay contacts have not 'welded' together due to arcing.

Precision temperature control

For very precise control of temperature, giving a stability of the order of 5 millidegrees, a toluene regulator and electronic relay may be used. The toluene regulator comprises a toluene filled bulb, a mercury tube, and a bimetal circuit enclosed in a housing at the top of the tube. As the temperature increases the mercury in the regulator tube rises and makes contact with an adjustable contact wire. The electronic relay is then operated and power to the water-bath heater and the bimetal unit heater is switched off. The bimetal strip then cools and the contact wire is arranged so that it is then lifted out of the mercury. The heaters are then switched on and this action is repeated. Since the bimetal strip takes longer to cool as the temperature increases, the ratio of heater ON to OFF time is progressively reduced. Fault finding in an electronic relay has already been described and when working on a toluene regulator it should be remembered that it is a precision device that should be handled with care. If the bimetal strip or heater fails to function then precision control will not be obtained and the temperature will fluctuate up and down to a greater extent than normal.

Thermocouples

Thermocouple junctions of dissimilar metals give an e.m.f linearly related to the temperature difference between a hot measuring junction and a cold reference junction. The introduction of connecting wires between the two junctions, introduces fresh junctions of dissimilar metals but providing their temperature remains the same as that of the original junction the overall e.m.f. is the same (law of intermediate metals). The reference junction may in fact be located within the measuring instrument.

Connecting wires have a temperature-sensitive resistance and thus, to avoid errors, as thick a gauge of wire as possible should be used (the thicker the wire the lower its resistance). Erratic and inaccurate results are obtained if the leads and contacts are dirty. Copper wires and contacts should be scraped to ensure absolute cleanliness.

Since thermocouples give a relative measurement, any variation in the temperature of the reference junction will cause a corresponding variation in the measured e.m.f. Potentiometric recorders, when used to record temperatures with thermocouples, incorporate a compensating resistance in the recorder measuring circuit to counteract the effect of reference-junction temperature changes.

For a laboratory experiment where a galvanometer is used to measure the thermocouple e.m.f., the cold reference junction can

GENERAL LABORATORY EQUIPMENT

be immersed in an ice/water mixture to maintain the temperature at 0°C. When low-resistance instruments with galvanometer movements are used for measuring, it is important to match the resistance of the leads (usually short) and thermocouple to the instrument, otherwise the galvanometer damping will not be correct. To avoid the introduction of spurious e.m.f.s, earth the circuit at one point only. Long connecting leads may be used with high-input-resistance potentiometric measuring circuits and particular care is necessary with earthing to avoid spurious signals. Spurious signals are usually due to stray electric currents flowing in the measuring circuit from external electric fields. Direct current stray can be eliminated by earthing at one point only, whereas alternating current stray requires connection of the instrument earth or shield terminal to either a positive or negative input terminal. The connection required depends on the source resistance, the amount of stray pick-up relative to earth, and on the location of the pick-up point in the measuring circuit. Reference should be made to the manufacturer's manual for the details relevant to a particular instrument.

A check for d.c. stray is shown in Figure 2.3(A) and infinite resistance on the testmeter indicates no d.c. path for stray signals. Less than 10 MΩ indicates a need to check insulation. To check for a.c. stray, first test the recorder by connecting a laboratory

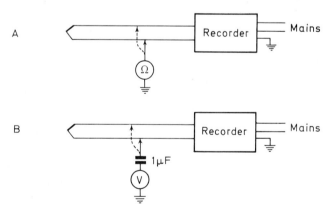

Figure 2.3. Checking thermocouple recorder circuits for stray signals.

test potentiometer to the input. This applies a known d.c. test voltage to the recorder and the operator can ensure that there are no internal faults. If the performance is satisfactory, note the sensitivity, calibration and recorder dead band. Compare these results with those obtained with the thermocouple connected. Any change indicates the presence of a.c. stray. This can be checked by measuring as shown in Figure 2.3(B) with a high-resistance a.c. voltmeter. The capacitor is necessary to block d.c. voltage levels which would affect the measurement.

When using thermocouples it should be remembered that there is a maximum recommended operating temperature for each type. These maximum temperatures are $371°C$ for copper constantan, $760°C$ for iron constantan, $1260°C$ for chromel alumel and $1538°C$ for platinum and platinum rhodium.

Other common temperature-measuring devices are the resistance thermometer and thermistor. The resistance thermometer typically comprises a platinum or nickel resistance element enclosed in a protective sheath. The resistance of the element changes with temperature. The thermistor comprises a small bead of semiconductor oxide within a protective covering. It is important that thermistor power dissipation is not exceeded or the semiconductor material will be affected.

Electronic thermometers
In electronic temperature-measuring circuits, the temperature-sensitive element (thermistor or resistance thermometer) usually forms one arm of a Wheatstone bridge. This circuit is illustrated in Figure 2.4 and typical maximum and minimum temperature conditions are shown in Figure 2.4(A) and (B). Instruments of this type normally have several ranges for temperature measurement and this is achieved by switching different resistances into one arm of the bridge and meter circuit. The bridge is balanced at the minimum temperature of each measuring range. As the resistance of the sensor changes with temperature, the bridge circuit becomes unbalanced and current flows through the temperature-calibrated meter. Electronic thermometers are often powered by dry batteries for portability and a discharged battery will be unable to provide sufficient voltage to give full-scale deflection on the meter

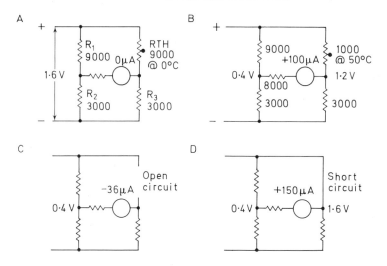

Figure 2.4. Electronic thermometer circuit operation.

at the maximum of the temperature range. Inaccurate readings are then obtained. To avoid this happening, the range switch often incorporates a check position which disconnects the sensor from the circuit and uses the meter to measure the state of the battery.

Typical faults occurring with this type of apparatus are the sensor or leads attached to it becoming open circuit or short circuit. The effects of these faults on circuit conditions are shown in Figures 2.4(C) and (D). With an open circuit sensor, the current flow through the meter is reversed (−ve). For a short circuit on the sensor circuit, excessive meter current flows in the normal direction (+ve) and the pointer will be hard over against the meter stop. Before looking for circuit faults, if say the meter reads backwards or the pointer is hard over to the right against the stop, check that the instrument is switched to the correct measuring range. The possibility of the knob of the range switch having been forced against the switch stop, causing it to slip round on its shaft, should not be overlooked. The knob should then be correctly positioned and the recessed screw securing it to the shaft tightened.

Dry batteries

On any portable apparatus dry batteries should not be left in the instrument if it is to remain unused for any length of time. This results in corrosion and may necessitate renewing not only the battery but also the battery holder. Leak-proof mercury and manganese alkaline types are advantageous in this respect and have a longer service life than a normal zinc-carbon battery. They have the disadvantage however that they are more expensive. It should be noted that mercury batteries may be used not only for the long service and storage life they give, but because they possess the most stable discharge voltage characteristic. Zinc-carbon batteries have the least-stable discharge voltage characteristic.

It may not always be possible to obtain locally, replacement mercury or manganese batteries, and in an emergency a near-equivalent zinc-carbon type can often be used. Battery holders are in fact of such dimensions that different makes and equivalent types of battery can often be fitted. Small differences in length are accommodated by compression of the coiled contact spring within the holder. The voltages of some near equivalent manganese alkaline and zinc-carbon batteries, listed in Table 2.1, are the same but mercury cells may be 1.4 V or 1.35 V, the latter voltage being indicated by a suffix R after the type number.

In the thermometer circuit described above, the voltage applied to the bridge should remain constant and for this reason the battery voltage may be passed through an adjustable resistance to the bridge circuit. The resistance is usually set by a screwdriver adjustment and enables allowance to be made for the fall of battery voltage with age. It must be reset whenever a new battery is fitted.

Null-balance measuring and control circuits

Precision measuring circuits often use a null-balance bridge circuit in which a sensitive centre-zero meter is connected across the bridge output. The resistance of one of the bridge arms is then varied to rebalance the circuit. The rebalancing variable resistance may be calibrated in temperature units (or scale numbers referred to a calibration chart) and is typically a multiturn helical potentiometer.

GENERAL LABORATORY EQUIPMENT

Table 2.1. Near-equivalent dry batteries up to $1\frac{1}{2}$ volts

	Zinc–Carbon	Mercury	Manganese alkaline
Type	SP 12	ZM-9	Mn-1500
Size (mm)	14.3 x 50.4	13.73 x 49.98	14.10 x 49.79
Voltage (V)	1.5	1.4	1.5
Max current drain (mA)	50	50	50
Type	SP 11	–	Mn-1400
Size (mm)	26.0 x 50.0		25.93 x 49.28
Voltage (V)	1.5		1.5
Max current drain (mA)	60		83
Type	SP 2	RM-42	Mn-1300
Size (mm)	34.0 x 61.1	30.33 x 60.78	33.35 x 60.39
Voltage (V)	1.5	1.4	1.5
Max current drain (mA)	100	250	250

In control applications the output from a temperature-sensitive bridge circuit may be applied to an amplifier which controls, for instance, the operation of a refrigeration system. This is done by a relay actuating the refrigeration unit motor or by controlling the flow of refrigerant through the opening or closing of a solenoid operated valve (mechanical valve actuated by passing a current through an operating coil). In this application, the bridge is balanced when the temperature is at the required value. An increase in temperature will decrease the sensor resistance, and the bridge then becomes unbalanced resulting in the circulation of refrigerant. When the temperature falls below a certain value the polarity of the bridge output reverses and the refrigeration system is switched off. Failure of this type of on and off control circuit to operate may result in the refrigerant flow being permanently on or off. If voltage can be measured at the connections of a solenoid valve which does not operate, then the valve is faulty. If no voltage can be measured when the valve should be energized, the fault may be in the amplifier or the resistance bridge. Thermistor sensors are often used in the bridge circuit and an open circuit in the sensor or its thin connecting leads results in a permanent output signal. Since the sensor indicates a low temperature by a high resistance, an open circuit may cause the solenoid valve to be permanently shut.

In a typical electronic control circuit for a heating load, a calibrated potentiometer controls via an electronic trigger circuit, the point on the applied alternating voltage waveform at which S.C.R. power controllers are turned on. With an alternating voltage applied to the S.C.R. anode-cathode circuit, conduction occurs during the portion of the waveform cycle when the anode is positive, providing the gate potential exceeds a certain value. By feeding the gate connection with a signal, the timing or phase of which can be varied with respect to the anode-cathode voltage waveform, the point on the waveform when conduction or turn-on occurs, can be controlled. Two S.C.R.s are often used in a back to back arrangement so that one conducts during the positive half cycle of the applied waveform and the other during the negative half cycle.

For proportional control, the temperature approaches its final value at a slower rate than it had initially, when the difference between the actual and required temperature was large. Many proportional-control heating circuits use a resistance bridge, incorporating the temperature setting potentiometer and a thermistor which senses the temperature of the controlled load. In low-power circuits, the thermistor may be heated by a small heating element through which the load current passes. The output from the bridge is amplified and applied to the S.C.R. power control stage. The temperature of the monitoring feedback thermistor increases as the controlled temperature nears its final value and its resistance decreases. This decrease in resistance changes the bridge output and the amplifier operates so that the S.C.R. cuts off earlier in the cycle of its applied voltage. In this way the ratio of ON time to OFF time is reduced as the required temperature is approached.

If no heat is produced in the load, the fault may be in the supply, the heating element or the electronic control circuit. Details of course depend on the circuit configuration used.

GENERAL LABORATORY EQUIPMENT

Generally the thermistor and temperature-control potentiometer are in series on one side of the bridge circuit and if they are open-circuit this results in a permanent offset voltage being fed to the amplifier input. A typical fault-finding procedure for failure to heat the load would be

(1) Check the fuses and continuity of the heater element with the power off. If the element is continuous and the mains supply is available, then test the electronic circuits.
(2) Check whether the input voltage to the amplifier varies when the temperature control is varied. If it does not, check the bridge circuit. Should no voltage be measurable at the amplifier input, check the d.c. power supply lines.
(3) With the above tests satisfactory, check the amplifier and S.C.R. trigger circuit components. If everything appears to be in order then the fault is probably in the S.C.R. units.

The S.C.R.s and preceding trigger circuits are best checked by using an oscilloscope, as the pulse waveforms can be examined and the point of malfunction established. A double-beam oscilloscope is useful for this, as one trace can show the S.C.R. anode waveform and the other the triggering signal. This testing would of course normally be done by an electronic technician. If maximum heat is continuously produced in the load irrespective of the position of the control potentiometer, then the S.C.R.s or the trigger control circuit are faulty. Failure to regulate to the required accuracy may be due to the monitoring thermistor resistance not changing with the temperature of the load.

ELECTRIC MOTORS AND SPEED CONTROL

Small-motor-driven apparatus

Laboratory bench apparatus utilizing small electric motors includes blenders, magnetic stirrers, liquid stirrers, pumps, whirli-mixers, shakers, viscometers, centrifuges, etc.

Various types of motor are used and the most common are small fractional-horse-power d.c. motors, single-phase a.c. induction motors, synchronous motors, shaded pole motors, and a.c./d.c. series motors. We can conveniently divide these into two groups (1) commutator motors and (2) induction motors. The universal a.c./d.c. series motor shown in Figure 2.5(A), utilizes a fixed-stator field-winding connected in series with a rotating armature. The field windings produce the magnetic field in which the armature current carrying conductors are situated. The electrical supply is connected to the armature through a commutator and carbon brushes. The a.c. repulsion motor is similar in construction to the series motor except that the mains supply is connected to the stator winding. The armature winding is connected to a commutator, the brushes being short circuited. The connections are shown in Figure 2.5(B). On some types there is a centrifugally operated commutator shorting device and brush lifting gear. These devices come into operation when the rotor has accelerated to a certain speed and cause the motor to run as a squirrel cage induction motor.

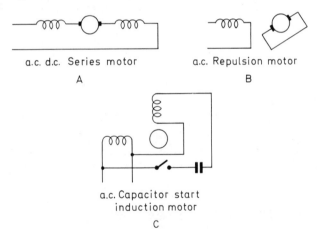

a.c. d.c. Series motor
A

a.c. Repulsion motor
B

a.c. Capacitor start induction motor
C

Figure 2.5. Single-phase a.c. fractional-horse-power motors.

Single-phase induction motors (connected between the live and neutral of the supply) consist of a squirrel-cage-constructed rotor and fixed stator windings. There is no brushgear and, except for certain small motors such as shaded pole types, they are not self starting. A rotating electromagnetic field is necessary for starting induction motors and in single-phase motors this is achieved by using an auxiliary starting winding and capacitor. The starting winding is spaced 90 electrical degrees from the main running winding and to produce the rotating field there must be a phase difference of $90°$ between the currents in the two windings. This is obtained by connecting a capacitor in series with the starting winding. The basic circuit is shown in Figure 2.5(C). The starting circuit is automatically opened by a centrifugal switch when the rotor has nearly reached its final speed. Synchronous motors are used where a constant known speed is required, as the speed of this type of motor is locked to the frequency of the a.c. supply.

In more powerful a.c. motors, which require a three-phase supply, the rotating field is produced by the windings being spaced electrically $120°$ apart.

Maintenance and testing of motors

Commutator motors require more maintenance than induction motors and for this reason the care of commutators is dealt with separately. The centrifugally operated switches used on repulsion and capacitor start-induction motors should be kept clean and well lubricated in order to maintain reliable operation. If the power supply fuses to a motor circuit blow, the motor can be checked for faulty insulation as follows. After switching off and disconnecting from the mains, disconnect the motor power leads from the switching and control circuit. Check the insulation resistance by connecting an insulation tester (megger) between one of the motor connections and the frame of the machine (earth). A reading greater than 1 megohm should be obtained for fractional-horse-power motors. If a lower reading is obtained, the motor should be replaced or repaired. An insulation breakdown on a complete motor assembly using a commutator and brushes may be due to 'tracking' of the supply voltage across the brush gear to earth due to an accumulation of dirt and carbon dust. This can be prevented by regular blowing out and cleaning. On universal commutator motors it is possible to ascertain whether the fault is in the armature or field winding, after isolating the two by removing the brushes from their holders and testing the armature and field separately. On capacitor induction motors, failure to start may be due to the capacitor and, when testing the capacitor, running and starting windings should be separated. This type of motor often incorporates a thermal cut out which disconnects the motor supply when the motor temperature rises due to an overload.

Small motors often have sealed bearings and require no regular lubrication. This is not always the case however and the manufacturer's manual should be consulted for details. Some motor bearings require grease and others light oil. An unusual noise, vibration or rough running indicates faulty bearings. This can be checked by attempting to turn the motor shaft by hand (with the power off). It should be free to rotate easily and smoothly. If the motor shaft cannot be turned by hand, then the bearings have 'seized'. The motor should then be repaired by an electrician, although in an emergency an attempt can be made to free the motor by working in light oil to the bearing. After allowing time for the oil to soak in, the motor shaft is worked backwards and forwards by hand until it can be made to turn easily.

Care and maintenance of commutators

Excessive sparking at motor brushes may be due to an uneven distribution of voltage within the armature due to a short or open circuit in the winding. Other causes are badly adjusted brushes, a brush jammed in its holder, and high mica insulation between the commutator segments causing jumping and chattering of the brushes. An armature can be tested for internal faults by applying a low d.c. voltage from a battery across brushes $180°$ apart (any other brushes should be removed), and measuring the voltage drop between adjacent commutator segments. On a good armature, all the voltage readings should be approximately the same. An unusual reading indicates an open circuit while a short circuit gives zero voltage.

It is important to use the correct grade of brush supplied by the manufacturer as the wrong grade will give rise to wear and sparking. Brushes for a.c. commutator motors are generally of higher resistance than those used on d.c. motors and the use of a soft-grade brush where a hard grade should be used will give excessive brush wear. It should also be remembered that different brushes need different spring tensions for correct operation. When new brushes are fitted, the motor should be run for a time to 'bed in' the brushes to the curve of the commutator.

High mica insulation can be removed, after removing the armature from the motor, by using a fine hack-saw blade. A normal commutator in use will have a yellow-brown colour while a bright copper pink is seen on a new or reground commutator. The use of glasspaper for commutator cleaning is not recommended although it may be necessary if the commutator is burnt. The electrical shop may 'skim up' a bad commutator in a lathe.

Electrical speed control of motors

The simplest speed-control circuit, applicable to small commutator motors, consists of a variable wire-wound resistance (rheostat) which controls the voltage applied to the motor. This type of circuit is typically used in controlling the speed of small stirrers up to for instance 2800 rev./min. Wear on the rheostat, where the moving contact crosses over the track of the resistance winding, may result in failure.

Blenders which require high speed and high power may utilize 1½ horse-power motors running at speeds up to 19 000 rev./min. Several speeds are obtained by selecting tapped windings on the motor or by using a variable transformer. Care should be taken not to overload blenders as they are designed only for intermittent operation. Running continuously for long periods of time and cutting very fibrous tissue without premincing can result in failure.

Electronic control of motor speed

A simple single S.C.R. a.c. series motor speed control circuit is shown in Figure 2.6. During the positive half cycle of the supply, capacitor Cl is positively charged from the zener diode through variable resistance R1, and when the S.C.R. gate current flowing through diode D3 is sufficiently positive the S.C.R. conducts. Speed control is obtained through controlling the time for the voltage on Cl to build up, by varying the resistance in the capacitor charging circuit. With a low resistance, the voltage rises quickly and the S.C.R. fires early in the positive half cycle. With a high resistance, the voltage rises slowly and the S.C.R. fires later in the cycle. In the negative supply half cycle, diode D2 becomes forward biased and diode D3 reverse biased.

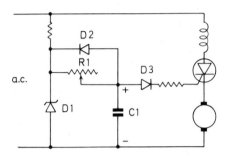

Figure 2.6. Silicon-controlled-rectifier motor speed control circuit.

Since the armature constitutes a conductor moving in a magnetic field, an e.m.f. known as the back-e.m.f. is induced in it. This back-e.m.f. acts with the capacitor voltage in such a way that an automatic speed regulation action is obtained. If the speed tends to rise, the back-e.m.f. increases and the gate signal is retarded causing the S.C.R. to fire later in the cycle. The speed is thus reduced.

With this type of circuit, if the motor will not run, after testing fuses, the motor brushes and the a.c. supply, check for open circuits in the speed-control resistance, diodes and resistances. Should the motor run at maximum speed with no control, the speed-control resistance R1 or diode D2 may be short-circuited. The capacitor C1 is then quickly charged to give a high positive gate input. If diode D2 is open circuit, the capacitor voltage will not discharge to zero on the negative half cycle of the supply voltage as it does in normal operation. Voltage then builds up on capacitor C1.

The Triac is a bidirectional semiconductor device and a power-control circuit suitable for both heating and motor loads is shown in Figure 2.7. On the positive half cycle of the supply, capacitor C1 is charged at a rate determined by the setting of potentiometer RV1. When the capacitor voltage rises to the trigger point of the bidirectional pulse diode D1, the diode 'fires' and a pulse is applied to the gate of the Triac. The Triac then conducts and current flows through the load. The point on the waveform at which the Triac fires depends on the setting of RV1. After firing, similar to the S.C.R., the Triac continues to conduct for the remainder of the half cycle until the alternating anode waveform passes through zero. This sequence is repeated for the negative half cycle of the applied waveform so that full wave-control is obtained.

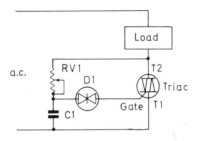

Figure 2.7. Triac power control circuit.

The application of this type of circuit to a laboratory bench centrifuge is described in the following section.

Laboratory bench centrifuges

A laboratory bench centrifuge spins biological materials at high speed for separation purposes. The tubes containing the liquid material are placed in a rotor which is driven by a series-wound electric motor. In the simplest type, the motor speed is varied by a wire-wound rheostat or a variable transformer which controls the voltage applied to the motor.

Maintenance consists principally of cleaning the commutator and replacing the brushes of the motor. Failure of the centrifuge to run, may be due to a blown fuse (due to a fault in the motor circuit), worn brushes or brushes stuck in the brush gear (brushes not making contact with motor commutator) or to an electrical fault in the speed control circuit (burnt out winding). On some centrifuges the centrifugation time is set on the calibrated dial of a timer, which switches off the power to the motor at the end of the preset time. Faulty timer contacts may prevent power being applied to the motor, or if the contacts stick or weld together, result in failure to switch off. Motors on these centrifuges usually have sealed bearings and require no lubrication.

Modern electronically-controlled bench centrifuges may use silicon-controlled rectifier or Triac circuits. A Triac circuit is used on the M.S.E. Super Minor bench centrifuge, the circuit basically operating on the principle described previously. If the centrifuge will not run, check the most likely causes such as faulty brushes or a blown fuse before suspecting faults in the electronic control circuit. A faulty Triac or pulse diode may be the cause of a malfunction in the electronic circuit, although it should not be forgotten that a fault in the pulse capacitor or its charging circuit will prevent the Triac conducting. Replacement of these semiconductor devices is best left to a trained engineer. Checking of resistances and capacitors can however easily be carried out by a laboratory technician. In power controllers, resistance-capacitance circuits are incorporated for radio-frequency interference suppression and a breakdown in these components can affect the operation of the speed control. On no account should any internal preset potentiometer in the circuit be altered.

Speed indication is obtained by measuring on a voltmeter (calibrated in rev./min) the output from a tachometer generator driven by the centrifuge motor. Some makes of centrifuge use brushes in the tachometer generator and erratic speed indication is a sign that the brush gear needs attention. In the Super Minor, an a.c. generator is used, the output being taken from the stator windings. There are therefore no brushes and the output is rectified and smoothed before being applied to the moving-coil meter. Oscillation of the meter needle is an indication that a.c. is being applied to it and the rectifier and smoothing capacitor should be checked. A preset potentiometer calibrates the meter to the centrifuge speed and should not be altered.

If the speed indicated is suspect or any components in the meter circuit have been replaced, recalibration is necessary. This is easily carried out by an electronic technician using the appropriate test equipment. Either a hand tachometer or stroboscope is required. The stroboscope method is preferred since this does not involve contact with the centrifuge rotor. Place a piece of tape or paint on the top of the rotor and then run the centrifuge up to a convenient speed with its lid open. The flashing light of the stroboscope, directed at the rotor, is then adjusted by the calibrated control until the tape or paint mark appears stationary. The stroboscope light is then flashing at a rate equal to the speed of the rotor. If the speed indicated does not agree with the stroboscope meter (calibrated in rev./min or pulse rate), the centrifuge preset calibration potentiometer requires adjustment. This method is only applicable to low-speed centrifuges in which accuracy is not important. Very-high-speed ultracentrifuges require more accurate setting up and a digital frequency meter may be required.

Sparking at centrifuge motor brushes will give rise to electrical interference on nearby sensitive apparatus and it is important to ensure that the suppressor circuit components, commutator and brush gear are in good condition. It should be remembered that it is not only sparking at motor brushes or relay contacts that produces interference, but unseen rapidly changing currents through semiconductor devices such as Triacs or S.C.R.s.

Electronic control circuits on high-speed and ultracentrifuges

A modern laboratory ultracentrifuge can spin samples to speeds of 75 000 rev./min at a controlled temperature. Such a machine will incorporate several control systems inter-connected by interlock and safety circuits, and usually include a rotary vacuum pump, diffusion pump, temperature-measuring circuit and temperature and speed control circuits. The construction of a Beckman-Spinco ultracentrifuge is shown in Figure 2.8.

Various electronic control circuits are used on ultracentrifuges and the fundamental operating principles are shown in Figure 2.9. The main drive motor, connected to the rotor drive shaft is also coupled to a small tachometer generator. The generator output voltage is fed back to the input of an amplifier and is opposite in polarity to the amplifier input from the speed-control potentiometer. Any difference in these two voltages results in an input signal to the amplifier and causes the power-control stage to adjust

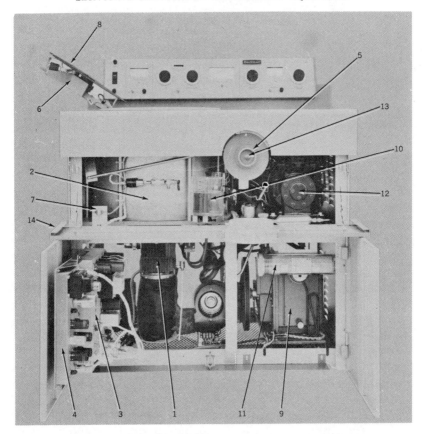

Figure 2.8. Construction of a Beckman L265 ultracentrifuge: 1. drive unit 2. rotor chamber 3. electronic speed control 4. electronic temperature control 5. chamber door control 6. rotor stabilizer 7. stabilizer fuse light 8. heat radiation shield 9. rotary vacuum pump 10. oil reservoir 11. drierite desiccant 12. refrigeration unit 13. vacuum pump button 14. drop-down door (courtesy of Beckman Instruments).

the motor speed to the required value. The generator voltage then balances the input from the speed-control potentiometer.

The drive unit may be cooled by water circulating in coils around the unit and the bearings lubricated by oil flowing by

GENERAL LABORATORY EQUIPMENT

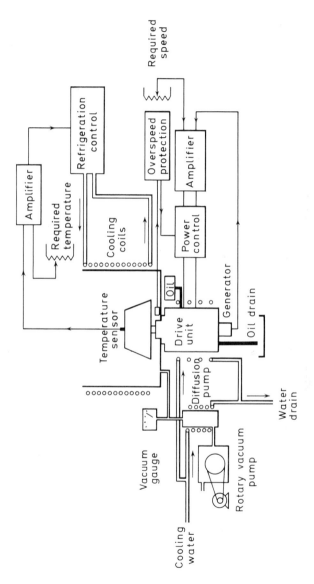

Figure 2.9. Ultracentrifuge control systems.

gravity feed from a reservoir or from a pump-driven recirculating oil system. Electrical interlocks are incorporated to prevent the drive motor running if the cooling water is not available (pressure-operated switch), if the reservoir oil level is low (micro-switch actuated by weight of oil) and if the chamber in which the centrifuge head rotates is not evacuated of air (pressure-switch and interlock with vacuum-pump motor circuit). As a safety precaution an overspeed circuit will remove power from the motor if the rotor speed exceeds the rated value by a small margin. Thus if the ultracentrifuge will not start, check that the interlock circuits are all correct before assuming that there is an electronic fault.

Control of the rotor temperature is usually obtained by using a null balance circuit to regulate the flow of refrigerant through coils wound around the centrifuge chamber or to switch on and off automatically the refrigeration unit compressor motor. Various types of temperature sensors are used including thermistor probes and infra-red sensors. Common difficulties are broken thermistor connecting leads and contamination of the 'window' of infra-red detectors by a film of oil.

Mains power supply and refrigeration compressor and drive motor fuses or overload circuit breakers are usually mounted on the control panel of the machine and can easily be checked. Amplifier and other control fuses may however be located internally. Transistor amplifiers on plug in printed circuit boards for ease of servicing are usual. Power-control stages often utilize magnetic amplifiers which are similar in appearance to a power transformer. The magnetic amplifier controls a large amount of a.c. current flowing in its secondary winding by a small change of d.c. current in the primary winding. These devices are very rugged and normally do not give trouble in service.

One of the problems in servicing large ultracentrifuges is that they are often situated against a wall or between benches for the convenience of the operator. This makes it difficult in many instances for the service engineer to check voltages at various points in the control and interlock circuits. This problem is however overcome in the M.S.E. Superspeed 75 which incorporates an internal test-panel accessible from the front of the

GENERAL LABORATORY EQUIPMENT

machine. The test-panel houses fuses for the mains voltage units in the ultracentrifuge and the speed control, overspeed protection and electronic power supply circuits. There are four, twelve-position test switches which operate in conjunction with neon lamps and a voltmeter. The first switch checks out the motor relay interlocks, the second the chamber lid interlocks and the vacuum system, the third the maximum rotor speed selector circuits and the fourth the electronic circuit supply voltages in conjunction with the voltmeter.

It must be stressed that scientists and laboratory technicians should confine themselves to first-line servicing only (i.e. checking oil levels, cooling water flow, motor brushes, fuses, etc). More detailed work should only be carried out by properly trained electronic/electrical technicians or service engineers. The replacement of components or plug in units of the electronic circuits often necessitates readjustment of internal preset potentiometers in association with special test equipment such as a digital frequency meter. This may be necessary on speed control and overspeed protection circuits for instance, and requires special knowledge which may only be available to the manufacturers service engineer. It is also unlikely that many laboratories will possess such expensive electronic test equipment.

A major source of breakdown on centrifuges using electromechanical relays are the electrical contacts. The motor power circuit is switched by a contactor with heavy-duty contacts which are actuated by an electromagnetic coil. These devices are rugged but do occasionally fail. Some machines are however prone to failure of the braking circuit. Without electrical braking, a heavily loaded centrifuge rotor running at high speed takes a considerable time to run down. At the end of the preset centrifugation time the contactor coil is de-energized and in order to brake the motor, the armature connections are automatically reversed with a high-wattage current-limiting resistance switched into circuit. This is shown in Figure 2.10, the motor power contactor being in the de-energized condition. The rotor speed is quickly reduced as the applied voltage is now trying to turn the motor in the opposite direction to normal drive due to the reversed armature connections When the speed has fallen to a few hundred rev./min a transistor-relay circuit R.L.B., normally energized from the tachometer generator output, is operated and the contact of R.L.B. opens the motor circuit.

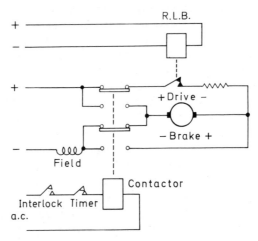

Figure 2.10. Centrifuge electrical-braking circuit.

Failure of this contact to open results in the centrifuge running down and then rotating in the opposite direction to normal. Complete failure of the braking circuit is usually due to the relay contact R.L.B. not making but may also be due to failure of the brake on/off selection switch or to the transistor operating the relay.

Bearings on high-speed motors are often troublesome and it should be remembered that they are centred and aligned to close tolerance limits. On ultracentrifuges the entire drive unit requires periodic replacement, while on lower-speed centrifuges (up to about 25 000 rev./min) the bearing housing assembly can be changed.

Care of centrifuge heads or rotors

Excessive vibration will occur if the liquid-filled tubes fitted into pockets or cups on the rotor, are not balanced to the required limits, or become unbalanced due to leakage during centrifugation.

Prolonged abuse of this nature may damage motor drive bearings and, on centrifuges with removable and inter-changeable heads, the resulting unbalanced force can cause the rotor to wobble and come off the drive shaft. This will not only damage the rotor but also the refrigeration coils and bowl of high speed and ultracentrifuges. For safety, the higher speed centrifuges have the bowl protected by an armour plated steel guard shield.

Rotors should always be carefully washed after use with a mild detergent followed by rinsing and drying. This is important in order to prevent corrosion occurring which may lead to failure of the rotor under the high stresses of centrifugation. While only very small cracks or fissures may appear on the outside surface, corrosive liquids may seep through and cause internal cavities which are only detectable by X-ray examination. A rotor which has disintegrated is shown in Figure 2.11.

Figure 2.11. A disintegrated high speed ultracentrifuge rotor.

VACUUM SYSTEMS

Vacuum pumps

Vacuum pumps often form an integral part of an instrument and since the vacuum system is usually interlocked with the electrical/electronic circuit, some notes on maintenance are included here for completeness.

Routine servicing of rotary vacuum pumps comprises topping up with oil (note on some pumps this must be done while the pump is running), cleaning the gauze filter in the vacuum line, and ensuring the tension in the drive belt between the electric motor and pump is correct. The drive belt tension can be altered by slackening the bolts fixing the motor to the base plate and adjusting the motor position.

If a condensable vapour (e.g. water vapour) is present in the vacuum system it is compressed to saturation point by the pump, and liquid enters the oil. This impairs the action of the oil in forming a seal and also its lubrication properties. The vacuum obtainable is also reduced as the liquid is in a closed system and circulates with the oil, evaporating into the vacuum chamber. A common method of preventing condensable vapour entering the oil is to use an air ballast valve. The vapour is then pumped directly to atmosphere, as the ballast valve introduces air which causes the pump exhaust valve to open before the vapour saturation point is reached. Ultracentrifuge vacuum systems are prone to contamination of the oil as the centrifuge chamber is encircled by refrigeration coils and moisture condenses on the chamber walls.

If the oil in the pump is contaminated, open the ballast valve fully and run the pump for several hours to remove the condensable vapour. In very bad cases it may be necessary to drain out the old oil and refill with fresh.

A high-vacuum pumping system usually comprises a forepump (rotary vacuum type) coupled to a vapour diffusion pump. For proper operation of the diffusion pump it is necessary to ensure that the cooling-water flow rate is correct and that the electrical heater for vaporizing the oil is functioning. Periodically the diffusion pump oil should be drained and the correct volume of fresh oil added. In order to avoid oxidation of the hot diffusion pump oil, it should be switched off, say 10 min before the rotary pump is stopped and air admitted to the chamber. This time-delay is often obtained automatically by means of a thermal delay relay. In systems where the evacuated chamber is also refrigerated, the rotary vacuum pump should be switched on before the refrigeration system. This is to prevent moisture from condensing on the walls of the chamber.

Pirani gauges

The Pirani gauge works on the principle that at low pressures the heat conductivity of a gas is linear with pressure. The Pirani gauge head comprises a platinum wire filament enclosed in a glass envelope with a connection to the vacuum system. When the pressure is very low there are fewer air molecules surrounding the wire, through which a small current is flowing, and its temperature rises and hence its resistance. An increase in pressure means that there are more molecules of air surrounding the wire, and its temperature falls slightly due to the resulting heat loss to the molecules. The change of resistance of the Pirani filament is proportional to the change in pressure and the wire forms one arm of a Wheatstone bridge measuring circuit. A second sealed Pirani element may be used in the bridge circuit so that the resistances of the two are compared. In order to avoid the effects of ambient temperature change on the calibration of the gauge circuit a compensating resistance may be wound on the gauge heads. Usually several Pirani gauge heads (containing the wire filaments) can be connected to the vacuum gauge measuring unit. The appropriate head is selected by switches. Failure of a Pirani gauge meter to read is usually due to the stabilized voltage supply to the bridge. A deflection hard left or right indicates an open circuit in one of the bridge arms, or that the user is operating on the wrong range. The most likely causes of trouble are switch contacts and contaminated gauge heads. To remove contaminants, the filament is 'flashed' (by passing current for 1 min) at a pressure *below* 10^{-4} Torr.

Tracing vacuum leaks

Leaks may be found in vacuum systems using a probe type leak detector. This type of detector may utilize an induction Tesla coil in which high-voltage high-frequency oscillations are produced by a transformer arranged in a tuned oscillator circuit, the transformer being supplied through an interrupter circuit. If the probe is held near a pin hole or very small crack in the vacuum system, a stream of sparks can be seen. In high vacuum systems the use of acetone for leak detection is however often more reliable. Acetone may be sprayed from a wash bottle and if a leak is present, acetone enters the system and the moisture causes the Pirani gauge meter-reading to change suddenly. A search jet of hydrogen or butane gas can also be used in a similar manner.

GENERAL ELECTRONIC APPARATUS

Variable transformers

Variable transformers (variacs) are basically single winding autotransformers with a movable carbon brush for the output tapping. They are used to control the supply of a.c. power to a variety of devices. Worn brushes give rise to arcing and may result in damage to the transformer winding. Dirt and dust on the contact track of the winding (on which the carbon brush makes contact) can also cause arcing which results in erratic operation. It is therefore important to clean the track regularly and to replace worn brushes. The track should be cleaned with a suitable solvent (e.g. carbon tetrachloride) and not by using an abrasive. The brush and track of a variable transformer in good condition are shown in Figure 2.12. A current overload will cause overheating and if severe will distort the winding turns and damage the insulation. The carbon brush may also disintegrate. Only the correct grade brushes should be used as the brush resistance limits to a safe value the circulating current flowing in the winding turns bridged by the brush.

Figure 2.12. Variable transformer winding and brush gear.

Constant-voltage transformers

Constant-voltage transformers are a special type of transformer, the tuned secondary circuit incorporating a capacitor unit. Circuits are shown in Figure 2.13. The output voltage is maintained constant to within a 1 per cent tolerance for a 10 to 15 per cent change in the input voltage. No maintenance is required although the capacitor can deteriorate if the unit becomes overheated due to inadequate natural ventilation. Constant-voltage transformers are self protecting in the event of an overload as the current is automatically limited to about 200 per cent of the normal full-load rating. Even with a short circuit in the load, the output

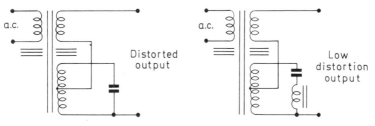

Figure 2.13. Constant-voltage transformers.

voltage falls to a very low value and the input power is limited to about 15 to 20 per cent of its normal value.

When checking the output voltage of a standard constant-voltage transformer, incorrect readings will be obtained if the correct type of meter (i.e. moving iron) is not used. A moving-coil d.c. meter incorporating a rectifier indicates a higher r.m.s. voltage (peak value of alternating voltage divided by $\sqrt{2}$) than is actually present due to the large amount of distortion present in the output waveform. Special types of constant-voltage transformers however have an almost sinusoidal output voltage (1 per cent distortion) and when testing them the above precaution need not be observed.

Unregulated d.c. power supplies

The basic circuit of a simple low-cost a.c./d.c. supply for educational purposes is shown in Figure 2.14. For safety, the mains voltage input is applied to a double-wound variable transformer and the low-voltage a.c. output is taken via an overload circuit breaker (trip) to terminals on the front panel. A d.c. output is obtained from a full-wave bridge rectifier.

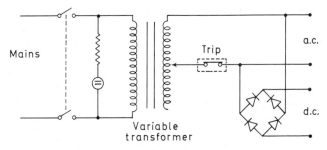

Figure 2.14. Unregulated power supply.

Circuit protection is given by the overload trip and if there is no output voltage, check whether the trip has operated. If it has, then ascertain and correct the cause (e.g. students wiring error) before resetting. If an a.c. output is obtainable but no d.c. voltage, then the rectifier is almost certainly faulty. This may be self evident by the unpleasant smell if a selenium metal rectifier is used.

GENERAL LABORATORY EQUIPMENT

Power-supply sections of instruments, supply d.c. voltages for the operation of transistors etc, and require some form of smoothing of the raw d.c. The operation of the full-wave rectifier circuit is shown in Figure 2.15, and the effect on the circuit waveforms of the resistance-capacitance smoothing filter can be seen.

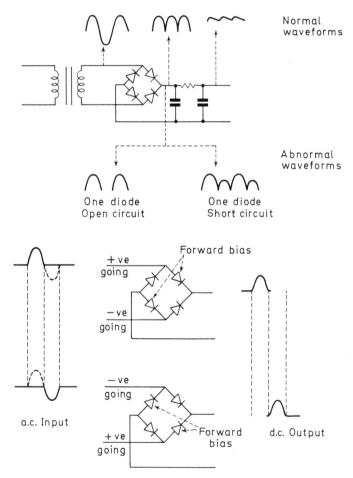

Figure 2.15. Full-wave rectifier circuit operation (see text for explanation of abnormal waveforms).

The degree of 'ripple' on the d.c. voltage level acceptable for proper operation of the electronic circuits is usually given in the manufacturer's service manual and can be checked using an oscilloscope. If the ripple is higher than normal, check that the rectifiers are functioning correctly before testing the filter components. The effect on the circuit of one rectifier going open circuit is shown in Figure 2.15.

If several rectifiers are connected in series in each arm of the bridge (as in some high voltage units) and one rectifier breaks down to give a short circuit, alternate half cycles of the rectified output may have different amplitudes. That is providing the remaining rectifiers can withstand the extra voltage across them. When replacing faulty rectifiers (or diodes) ensure that the replacements have the same or greater current rating and that they will withstand the circuit peak inverse voltage. The peak inverse voltages for various rectifier circuits are given in terms of the r.m.s. value of the transformer secondary voltage in Table 2.2.

Table 2.2. Rectifier ratings on single-phase rectification circuits
(Ed.c. = d.c. output voltage; Id.c. = d.c. output current;
E_T = transformer secondary r.m.s. voltage).

	Half Wave	Full Wave centre tap	Full Wave bridge
Peak inverse voltage rating of rectifier	3.14 Ed.c. 1.41 E_T	3.14 Ed.c. 2.82 E_T	1.57 Ed.c. 1.41 E_T
Average current rating of rectifier	Id.c.	0.5 Id.c.	0.5 Id.c.
Peak current rating of rectifier (resistive load)	3.14 Id.c.	1.57 Id.c.	1.57 Id.c.

Stabilized power supplies

Voltage supplies virtually independent of mains voltage changes and load currents are obtained using stabilizer circuits. The simplest circuits use only cold cathode valves or zener diodes, but more precise control is obtained by circuits in which a portion of the output voltage is compared with a reference voltage and any difference amplified and used to correct the output voltage level. The principle is shown in Figure 2.16. The amplifier output controls a power regulating transistor in series with the load, and the transistor's effective resistance is changed in order to counteract any change in the output voltage.

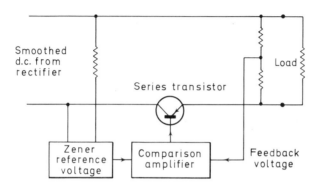

Figure 2.16. Operating principle of a stabilized power supply.

Stabilized power supplies may also give a constant current output. In this type, the load current flows through a low-ohmic-value resistance to develop a voltage which is compared with a reference voltage in the manner described above.

In a stabilized power supply circuit, if no output voltage is obtained, check fuses, the rectifier circuit and finally the series regulating transistor for an open circuit. Should the output voltage be at an abnormal value, check the reference voltage level with a voltmeter and the comparison amplifier for faulty components.

High-voltage power supplies

High d.c. voltages are usually obtained by using rectifiers and capacitors in special arrangements which give voltages that are multiples of the power supply transformer peak secondary voltage. These circuits are known as voltage doublers, triplers etc.

High-voltage units may rectify the a.c. mains supply if an appreciable amount of power is to be supplied to the load. Often however the current load is small (only a few milliamperes) and at a voltage of 1000 V this represents a load of perhaps 5 W. In such applications, an oscillator is used to provide a high frequency alternating voltage which is then coupled to the voltage multiplier circuit to produce a high-voltage d.c. output. The use of an oscillator to provide the alternating supply gives a high frequency ripple voltage which is easily smoothed by low-value capacitors.

High-voltage d.c. power supplies for physics laboratory demonstrations and experiments must be designed not only for reliability but for optimum safety. An example of this type is the Linstead S2 EHT power supply shown in Figure 2.17. A radio-frequency

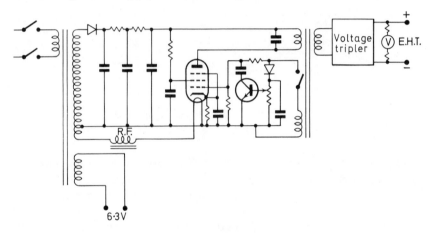

Figure 2.17. Educational EHT power supply (courtesy of Linstead Electronics).

phase-shift type valve oscillator produces fixed-frequency oscillations that are coupled to a voltage tripler circuit through the coupling between a coil in the valve anode circuit and the EHT coil. The principle of the oscillator is described in the section on Signal Generators. The voltage output is varied by a low-voltage potentiometer which controls, via a transistor, the voltage on the grid of the oscillator valve and thus the amplitude of oscillation.

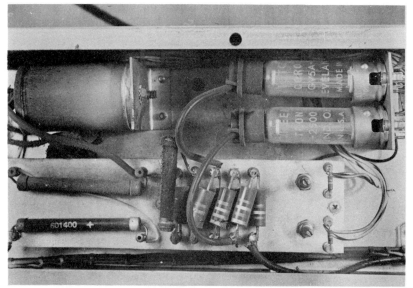

Figure 2.18. Attraction of dust particles to components in a high-voltage power supply.

The oscillator valve current is limited under external short-circuit conditions by grid and cathode biasing resistances. The oscillator cannot then maintain oscillation and the unit output current is small.

No high-voltage output will be obtained if the oscillator fails to function and the most likely cause is failure of the valve. Failure of the transistor or an open circuit in the voltage-control potentiometer will prevent the voltage from being set to the required value. On high-voltage power units it is important to keep the components clean and to avoid electrical leakage due to the accumulation of dust and dirt. Dust may be attracted electrostatically and this is shown in Figure 2.18 which shows part of the circuit of a photomultiplier EHT supply.

Electrophoresis power supplies

In the technique of zone electrophoresis, a mixture to be separated is applied as a spot on a paper strip. The paper is set up in an electrophoresis tank and a current is passed through the paper at high voltage. This separates the components of the mixture into a number of discrete zones or bands.

For voltages of a few hundred volts, both constant voltage and constant current power supplies are used. With a constant voltage applied, the current will change from its initial value due to the change in the resistance of the paper as it warms up.

Units operating at very high voltages (5 or 10 kV) are not stabilized and special safety precautions must be taken. Since the power dissipated may be several kilowatts this amount of heat must be removed from the tank and this is typically done by circulating cooling water in coils immersed in the tank. In a flat-bed arrangement the paper lies between metal clamping plates but insulated electrically from them by polythene sheeting. Water is then circulated through internal channels in the metal plates.

Safety interlocks are provided in the power-supply control circuit to cut off the power automatically when the apparatus cover or electrophoresis chamber door is opened. On some installations additional interlocks operate if the cooling water flow fails or the tank temperature exceeds a preset value. Toluene is used in some tanks and because of the fire hazard automatic CO_2 fire extinguisher systems are incorporated in large installations. They should be regularly checked by the manufacturer.

A typical single-phase power unit is shown in Figure 2.19 (high-power units often use a three-phase supply). The a.c. supply is fed to a variable transformer via a contactor and the variac output is then applied to a step-up transformer and voltage-

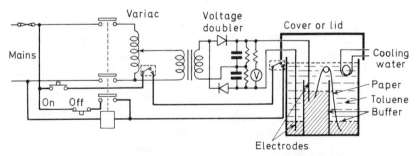

Figure 2.19. High-voltage electrophoresis apparatus.

doubling rectifier circuit. A microswitch is used in the coil circuit of the contactor for the lid safety interlock.

If no output voltage is obtained, check the fuses and the safety interlocks. If the chamber door or lid is not properly closed then the contactor cannot work and there is no mains voltage applied to the transformer. Often an additional microswitch is linked with the moving contact of the variac, so that power can only be switched on with the voltage control in the minimum voltage position. Ensure that the control is fully anticlockwise when switching on. Plug and socket connections and interconnecting cables between the tank and power unit are all possible sources of malfunction. Should the fuses have blown it is most likely due to a fault in the connecting leads to the tank or to an insulation breakdown on the high-voltage circuit.

The resistances across the output capacitors form a bleeder chain to ensure that the charge held on the capacitors leaks away quickly when the power is switched off. Always check that the capacitors are discharged before working on the internal circuit of the power unit. Remember there may be an open circuit in the bleeder chain. If while in use, arcing or sparking is heard from within the power unit, switch off at once. It is important to keep the high-voltage components clean and dry and free from dust in order to avoid leakage and tracking of the high voltage to earth. The potential dangers of high-power electrophoresis units cannot be overemphasized as the voltages are extemely lethal. They should only be serviced by trained technicians. Another potential hazard is the cooling-water supply and the condition of pipes, tubing and water connections should be regularly checked. Water and high-voltage electricity do not mix well together.

Signal generators

Square or sinusoidal waveforms are often required in teaching laboratories to produce audio and ultrasonic signals and by the electronics technician for examining circuits.

If we take a voltage amplifying stage, and feed back part of the output voltage to the input in such a way that the voltage fed back is electrically in phase with the input, then oscillations can be produced. Electronic sinusoidal oscillators utilize this principle and the 'positive' in phase feedback is obtained by inductive or capacitance coupling or the use of resistance–capacitance networks between the amplifier output and input. The Wein bridge oscillator uses a two stage amplifier connected between the input and output of a frequency-dependent bridge circuit. A constant amplitude output is obtained by the use of a temperature-sensitive resistance in a negative 'out-of-phase' feedback path in the amplifier.

The Wein bridge oscillator is often used in signal generators and is isolated from the load by a buffer amplifier stage. The output from the buffer amplifier is passed through a variable resistive voltage divider circuit (attenuator) to the generator output socket.

Block diagrams showing the sine and square wave modes of operation of the Linstead G2 educational signal generator are given as an example of this type of equipment in Figure 2.20. A valve Wein bridge oscillator is used with a transistor emitter-follower buffer amplifier. The emitter-follower transistor configuration matches the oscillator high-output impedance to the lower resistance of the load. It has approximately unity gain. A

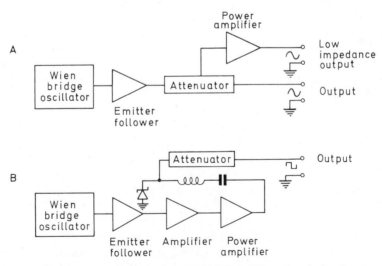

Figure 2.20. Modes of operation of an educational signal generator (courtesy of Linstead Electronics).

very-low-impedance sinusoidal wave output is also available through a power amplifying stage. In the square wave mode, the sinusoidal wave output from the oscillator is acted upon by the transistor amplifier and power-output stage with a zener diode to produce a square wave output. In both modes, switched resistances are used to cover the 10 to 100 kHz range in four decades in conjunction with continuously variable capacitors.

An output that fluctuates in amplitude may be due to a faulty thermistor or valve in the oscillator circuit. The absence of any output signal can be caused by failure of the oscillator valve, the emitter follower or to a faulty switch contact or open-circuit resistance in the attenuator. A distorted waveform will be obtained if the load resistance is too low and the oscillator may then become unstable. Educational equipment of this kind is usually very conservatively designed so that breakdowns due to component failures are rare.

The accuracy of the output signal should be periodically checked by connecting an oscilloscope across the outputs. It should be noted however that the 'squareness' of the square wave is greatly affected by the frequency response of the circuit to which it is connected. In fact use is made of this when testing amplifiers as will be seen later.

To measure the frequency of the output signal, adjust the switched time base of the oscilloscope to a value that gives one cycle of the waveform per division of the oscilloscope screen graticule. If one division represents 100 microseconds (μ sec) and is filled by one complete cycle, then the frequency is given by $1 \times (1\,000\,000/100)$ which equals 10 000 cycles per sec. When measuring, take care that any variable time base control is fully at the end of its travel in the CAL (for calibration) position. The waveform can be moved from left to right (or vice versa) by using the oscilloscope shift control.

An oscilloscope should be used for fault location. For example, if no signal is obtained from the low impedance output terminals, work back through the power amplifier and emitter follower to the oscillator valve. If for instance, a waveform can be measured at the oscillator output but not at the emitter follower output, the fault lies in the latter section.

Audiofrequency amplifiers

A simplified circuit of an audiofrequency amplifier is shown in Figure 2.21. The input signal is amplified and applied to a power driving stage which is coupled by a transformer to a push-pull output amplifier. Push-pull amplifiers are usually operated so that the transistors are biased almost to cut off of collector current when there is no input signal. This mode of operation is known as class B and gives a large power output. The signal amplification stages are biased so that the transistor is operated in the centre of its characteristic. This is known as class A operation and gives linear amplification with low distortion.

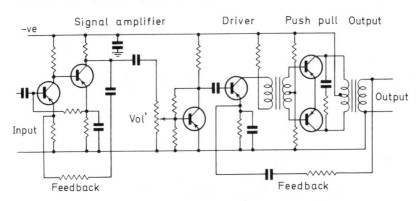

Figure 2.21. Basic audiofrequency amplifier circuit.

During the positive half cycle of the power amplifier driving stage output, one of the push-pull transistors conducts and the other is cut off. On the negative half cycle, the previously conducting transistor is cut off and the cut off transistor conducts. The push-pull transistor outputs are combined in the output transformer to give an amplified version of the signal.

Typical amplifier faults are (1) no output (2) distorted output (3) low output volume [often combined with (2)] and (4) hum on output.

If no output at all is obtained, after ensuring that the fuses are intact, the circuit should be checked in the following sequence: (1) d.c. power-supply voltages and (2) voltage and signal levels of

GENERAL LABORATORY EQUIPMENT

amplifier stages. If a sine-wave generator and oscilloscope are available, a suitable test-signal can be fed in (say 100 mV at 1 kHz) and the point in the circuit when the oscilloscope trace of the amplified signal disappears, indicates the faulty stage. The fault may be due to an open circuit in one of the amplifying stages or in the interstage coupling. Should there be no d.c. voltages measurable on the power supply lines check for an open-circuit resistance in the smoothing network or for an open-circuit transformer winding or rectifier.

If the fuses have blown, a possible cause is breakdown of one of the smoothing capacitors. The short-circuit current flowing in the very brief time before the fuse blows may have damaged the rectifiers, and they should be checked for satisfactory forward and reverse resistance. Excessive current drain on the power supply may also blow the fuses and may be due to a fault anywhere within the amplifier. Systematic resistance testing of components (with the amplifier switched off and disconnected from the mains) is then necessary. Physical signs of overheating, etc., may however assist in locating the fault.

Distortion can be due to one of the transistor stages operating in a non-linear manner and the point in the circuit at which distortion first occurs, can be found by monitoring the output of each stage with an oscilloscope while a signal is fed in at the amplifier input from a signal generator.

In amplifiers, part of the output voltage is fed back to the input so that the input and voltage feedback are opposite in phase. This negative feedback makes the amplifier gain and frequency response dependent on the feedback circuit passive components (e.g. resistances) and independent of variations in transistor parameters. Negative feedback also reduces distortion and this is the reason for the feedback loop in the class B amplifier output stage.

The frequency response of an amplifier (i.e. its gain over a frequency range) can be ascertained by feeding in square waves from a signal generator and monitoring the amplifier with an oscilloscope. A square wave can be considered as a fundamental-frequency sine wave plus an infinite series of odd harmonic sine waves. A perfect amplifier would not affect the shape of the square wave, but in practice, amplifier defects show up in the shape of the waveform obtained at the output. If low-frequency components of the applied square wave are not present at the output in the correct amounts and phase relationships, the flat top of the square wave is distorted. Poor high-frequency response gives curved leading edges, and spikes on the top of the leading edge of the wave indicate excessive high frequency response or damped oscillation. No practical amplifier is perfect and some distortion at frequencies below say 100 Hz and above 10 kHz is to be expected. Some typical oscilloscope displays obtained on square wave testing an amplifier are shown in Figure 2.22.

Hum or unwanted ripple voltage superimposed on the output may be due to 'pick up' from mains-frequency wiring, failure of

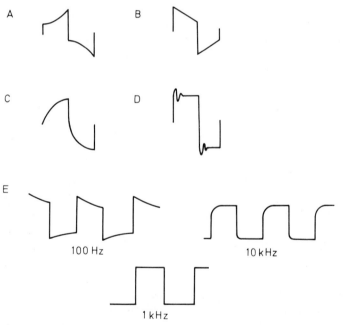

Figure 2.22. Square-wave testing an amplifier.
A. Poor low-frequency response   B. Excessive low-frequency response
C. Poor high-frequency response   D. Excessive high-frequency response
    E. Normal response at low-, mid- and high-frequency

the rectifier circuit smoothing capacitors or to poor earthing arrangements. Ensure that all screened leads have the screens connected together and earthed at one point.

On class B power amplifiers, if the load impedance (opposition to flow of alternating current) falls below a minimum value, the output transistor current increases to such an extent that the transistor is destroyed. It is therefore particularly important not to short circuit the output socket accidentally when plugging in the external load (e.g. loudspeaker). Sophisticated equipment will include overload protection devices such as thermal overload cut-outs or overcurrent protection circuits.

Timers

The simplest timer, typically available in ranges of 0-10 min to 0-5 h, is the spring-driven manually-operated type. The clockwork spring is automatically wound and electrical contacts made on setting the pointer to the required time. At the end of the time period the electrical contacts open.

A much wider range of time periods, typically 0-1 min to 3-60 h, is obtained with a synchronous motor-driven timer. Setting the pointer to the required time operates change-over contacts and starts a synchronous motor unit. On completion of the set time, the motor is switched off and the contacts revert to their original state. Operation of such a timer connected to an external load is shown in Figure 2.23. By suitable wiring the timer can be used to switch a load on or off for preset time periods.

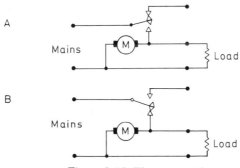

Figure 2.23. Timer operation.

On some timers, physical damage occurs to the mechanism if the pointer is not set to the required time in a particular direction (usually marked by an arrow). In use, most troubles arise from deterioration of the electrical contacts. Some timers use a small microswitch unit which is actuated by a cam driven by the synchronous motor. If the cam is not tight on the motor shaft, slipping occurs and a fault condition results. On such timers, the synchronous motor and a reduction gearbox are combined in one unit, and since the speed of the output shaft may be slow (e.g. 1/10 of a rev./min), it is sometimes difficult to tell whether the unit is working.

Electronic timers are used for accurate control of short time delay periods. The principle commonly used is to charge a stable low-leakage capacitor through a resistance. When the capacitor voltage builds up to a certain value (dependent on the circuit time constant given by resistance in ohms multiplied by capacitance in farads) an electronic element such as a valve, relay or transistor is operated. Alternatively the capacitor may be charged and the time delay period initiated by operating a switch which discharges the capacitor through a relay operating coil.

Apart from contact troubles, incorrect timing results if the capacitor does not hold its charge (i.e. becomes leaky). Complete failure to operate can occur if the resistance or capacitance go open circuit, if the initiating switch does not make contact, or the electronic unit fails.

For very precise timing measurements, an oscillator and electronic counting circuit are used. This type of equipment is described in Chapter 6.

A unijunction transistor is often used with electrical resistance-capacitance timing circuits. It is a special semiconductor device, the controlling P-type emitter connection being made at a point along an N-type base of bar construction. One end of the base is normally connected to the positive side of the d.c. supply and the other via a load resistance to the common line. A current pulse is delivered through the load and may be used to trigger an S.C.R., when the emitter potential reaches a certain value.

3 Electrochemical Measuring and Analytical Equipment

ph ELECTRODES AND METERS

pH measurement
The pH of a liquid is a measure of its acidity or alkalinity, which is related to the concentration of hydrogen ions in the solution. A system of numbers is used to indicate the pH value which is defined as the negative logarithm of the hydrogen-ion concentration.

The measurement of pH is carried out using electrodes which are immersed in the liquid or semi-solid being measured. The electrodes are connected to a pH meter which is a high-input-impedance d.c. amplifier with a meter read-out calibrated in pH units and millivolts. Two electrodes are used, a very-high-resistance glass measuring element and a reference element, the electrical output of the electrode pair being proportioned to pH. Single combined electrodes which incorporate both the measuring and reference elements in one assembly are also available. It is important to note that glass electrode elements give zero electrical output at a particular pH value. Typical zero e.m.f. (E_0) values are 2, 6.5 and 7 pH. When replacing electrodes it is important to check that they have the correct zero e.m.f. value for the pH meter electronic circuit.

pH and temperature
The pH value of a solution varies with temperature and when making up buffer test solutions, for standardizing or checking a pH meter, this must be taken into account. If a solution is warmed to dissolve buffer tablets then either the temperature must be

measured with a thermometer or the solution allowed to cool before measurements are made. Failure to do this will give incorrect results. It should also be remembered that the slope of the pH–e.m.f. curve of an electrode pair changes with temperature. This necessitates changing the sensitivity of the pH meter amplifier either manually or automatically. In automatic temperature compensation circuits a resistance thermometer is immersed in the solution under test.

Care of electrodes

Faults reported on pH meters are often not due to the electronic circuit but to the condition of the electrodes. A contaminated bulb membrane on a glass electrode gives a sluggish response and reduced pH range while a crack in the membrane results in a constant reading on the meter irrespective of the pH of the solution. A blocked porous plug at the base of the reference electrode gives a similar effect. The cleaning procedure for contaminated glass electrodes depends on the nature of the contamination. Grease can be removed by using acetone, protein by soaking in 0.1M-hydrochloric acid with pepsin, and inorganic deposits by acids. Electrodes are reactivated after cleaning usually by soaking for 24 h in 0.1M-hydrochloric acid followed by 24 h in distilled water.

New glass electrodes also require activation before use, although it should be noted that some types of single and combined electrodes should be soaked in potassium chloride or distilled water depending on the method of dispatch and the glass membrane used in their construction. The reference electrode should normally be 2/3 filled with saturated potassium chloride. Before activating and using new electrodes, remove rubber teats in which the glass membrane may be soaking, and protective caps or tape from the potassium chloride filling arm or ceramic liquid junction. Never allow electrodes in routine use to become dry. Some typical faults due to electrodes are summarized in Table 3.1.

Simple pH meters

A simplified diagram of the Pye 78 pH-meter, which is a simple d.c. amplifier instrument, is shown in Figure 3.1. Only the circuit

ELECTROCHEMICAL MEASURING AND ANALYTICAL EQUIPMENT

Table 3.1. Typical faults due to pH electrodes

Symptom	Probable cause	Action
Erratic readings	Bad electrical connection between electrode lead plug and meter socket	Clean or replace
Constant pH reading on meter	Glass electrode membrane cracked Blocked porous plug on reference electrode	Replace
Sluggish response	Contaminated electrodes	Clean or replace
Insufficient range	Contaminated glass electrode Unsaturated KCl in reference electrode	Clean or replace Top up with fresh saturated KCl

for pH measurement over the range 0-14 pH is shown although in fact measurements can be made on a 3.5 to 10.5 pH range and from -175 to $+175$ mV with an offset bias from -1000 mV to $+1000$ mV in 200 mV steps. The basic circuit is typical of many valve-operated pH-meters which are still widely used although the latest instruments available are of course transistorized.

The amplifier section comprises a pair of matched ME1400 valves. The electrode output is applied to the grid of the input valve while the grid of the second valve is fed from a potential divider chain incorporating a 'set 7' or fine zero control. The high impedance output of these amplifier valves is matched to the low-resistance microammeter by a cathode-follower-connected double-triode ECC 81 valve. The circuit is set up with zero electrical input (i.e. with the range switch on the 7 pH position) and the zero or set 7 control is adjusted until the output centre zero meter reads exactly mid-scale. It should be noted that the current output is 0 μA at 7 pH, $+700$ μA at 14 pH and -700 μA at 0 pH.

The electrodes connected to the pH meter have a very high electrical resistance and the amplifier input characteristics are

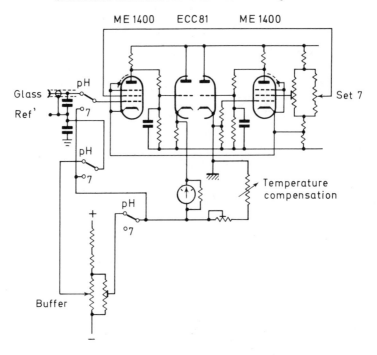

Figure 3.1. Simplified circuit of Pye 78 pH meter.

represented by a resistance of $10^{12}\Omega$ in parallel with a capacitance of 2000 pF. The current flowing in the input circuit is typically only 5×10^{-12} A. The effects of leakage, moisture and valve grid current are therefore important. Grid current flows in a valve grid circuit due to a positive grid electrode collecting electrons emitted from the valve cathode or to the production of electrons due to the action of light.

The latter source of grid current is eliminated by coating the valve envelope with opaque paint and operating the heaters at a low voltage to reduce the heater glow normally obtained. Valves designed for this type of operation are known as electrometer valves. The insulation of the input circuit to earth is important as the leakage currents which flow give rise to incorrect readings. For this reason the electrometer valve should be in a dry atmosphere, and in order to avoid spurious erratic readings the pH meter should not be switched off at the wall socket between measurements. The insulation resistance of the electrode leads (normally co-axial with polyethylene or Teflon insulation) is also important.

Stable amplification is achieved by the use of negative feedback. For automatic temperature compensation, the resistance thermometer forms part of the feedback circuit and varies the amplifier gain according to temperature. An open circuit thermometer causes a break in the amplifier circuit and the output meter will not respond correctly to pH changes. A fault-finding chart for this type of instrument is given in Table 3.2.

Table 3.2(a) pH meter fault-finding guide

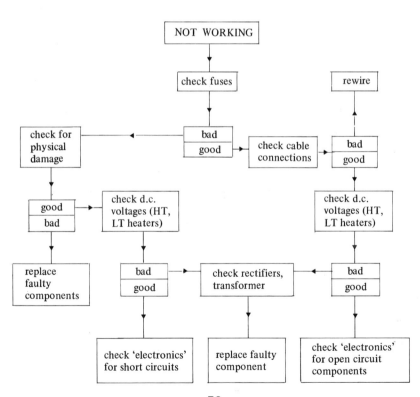

Table 3.2(b)

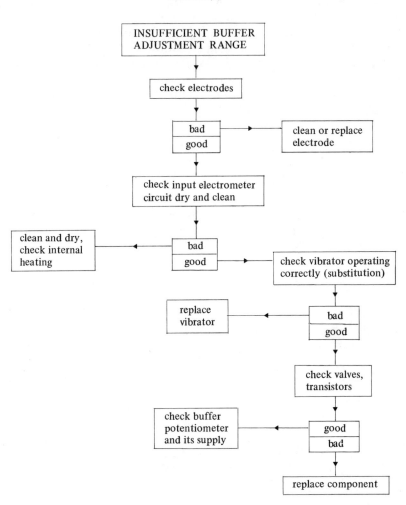

ELECTROCHEMICAL MEASURING AND ANALYTICAL EQUIPMENT

Table 3.2(c)

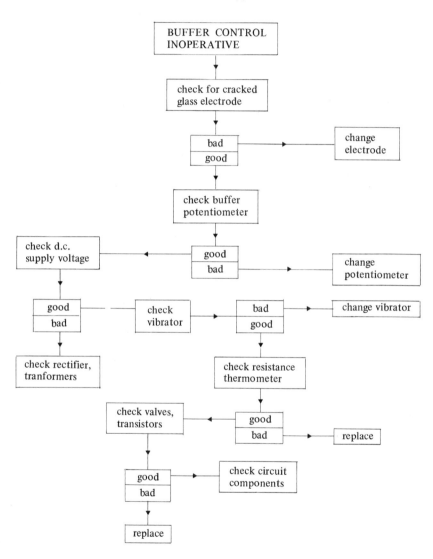

Chopper and vibrating capacitor pH meters

Direct-coupled amplifiers are particularly subject to drift (change of output when input constant) and stabilized-power-supply circuits are necessary. In the Pye 78 pH meter, the valve H.T. supply is stabilized by cold-cathode neon tubes and the ME1400 heater supplies are stabilized by a two-zener diode and transistor regulator circuit. In order to overcome the problems of d.c. amplification in many types of pH meter the meter input signal from the electrodes is converted to an alternative signal. This is done by using a converter which may be an electro-mechanical chopper (vibrator) vibrating capacitor, or photo-conductive modulator.

The electro-mechanical chopper consists basically of an a.c. energized electromagnetic coil, operating a set of change-over electrical contacts. The centre moving contact alternately makes and breaks with the two fixed contacts. The vibrator is connected between the input glass electrode socket and a centre tapped transformer, so that an alternating signal proportional to the amplitude of the d.c. input is obtained from the transformer secondary winding. In practice vibrators are plug-in units as shown in Figure 3.2. Normally the vibrations from the unit can be felt with the finger tips if the unit is operating correctly. Vibrators are sealed units and a new vibrator should be fitted if a faulty unit is found. As with any electrical contacts, deterioration occurs after a period of use, and this problem is overcome by the use of photoconductive and vibrating capacitor converters. The Pye 290 pH meter is an example of a modern instrument which uses a photomodulator converter. A simplified circuit is shown in Figure 3.3. The photo-sensitive resistor has a high electrical resistance when not exposed to light and a low resistance when exposed. Thus by focusing the light from a neon lamp, which is flashing on and off, on to the resistance, it is made to act as a switch alternately opening and closing. This gives rise to a 'chopped' signal which is passed through a filter circuit to the input stage of the a.c. amplifier. The amplifier is all solid state (i.e. transistorized) and the input stage is a field effect transistor (F.E.T.) which has a very high input resistance and can be

ELECTROCHEMICAL MEASURING AND ANALYTICAL EQUIPMENT

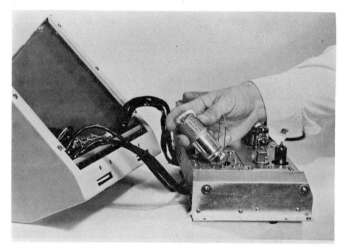

Figure 3.2. Fitting a new vibrator unit to a pH meter amplifier removed from its case.

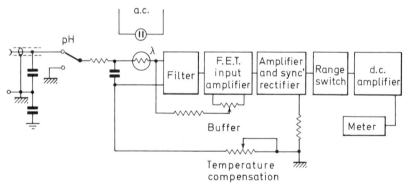

Figure 3.3. Diagram of Pye 290 pH meter.

considered the semiconductor equivalent of the electrometer valve. After a.c. amplification the signal is converted back to d.c. (by a special rectifier circuit working in synchronism with the converter or modulator unit), further amplified and applied to the output meter.

The input circuit is very sensitive and is enclosed in a metal case for shielding. This circuit includes very high-value resistances (e.g. 5 kMΩ) and if the instrument is subject to rapid changes in temperature, or high humidity, drifting and erratic behaviour will result. Erratic flicking of the output meter needle can be due to interference from electrical and electronic switching circuits (particularly commutator motors and S.C.R. power controllers). The best remedy is to remove the pH meter (or the offending apparatus) to another location. Interference suppressors can be fitted to the mains lead but complete elimination of interference is sometimes difficult.

If the pH meter is suspected of giving incorrect readings, check first that the mechanical zero on the output meter is correct. If the meter needle is not on zero, then with the instrument switched off, adjust carefully by a small amount the slotted screw in the centre of the meter.

The electrical zero (7 pH point) can be tested by removing the electrodes, and short-circuiting the input terminals. With the range switch on 0–1.4 and the function switch alternately switched between mV+ and mV− there should be no alteration in the meter reading. If there is a difference, the internal amplifier preset zero potentiometer may need adjustment.

To check the amplifier sensitivity, a laboratory potentiometer accurate to 0.05 per cent is necessary. With the range switch on 0–14 and mV+ selected on the function switch, check the meter zero or 7 pH point. Next apply 1400 mV from the potentiometer and check that the meter reads correctly. If it does not, then the internal preset sensitivity potentiometer may need adjustment. A variable mV source (consisting of a switched potential divider chain fed from a battery) can be used for a functional check on a pH meter although for calibration an accurate potentiometer source should be used. The above checks, while described using the Pye 290 pH meter as an example, can of course be applied in principle to any pH meter. For a proper understanding of the operating controls of a pH meter it is important to read the instrument instruction manual.

Should it appear that there is a fault within the amplifier of a pH meter, measure first the power-supply voltage lines (e.g. in the Pye 290 +15 V and −16 V). If these are satisfactory then, after looking for visual signs, measure the voltage at the various points indicated on the circuit diagram. An abnormal reading indicates the location of the fault. It should be noted that replacement of some components should be carried out by a service engineer or trained technician as certain resistances for example are often selected on test while the replacement of a zener voltage reference diode may necessitate circuit adjustments. This is normally stated on the circuit diagram or parts list. The procedure to be followed however may only be known to the manufacturer's service engineer or to a customer's technician who has been trained in the manufacturer's service training school.

In order to expand the scale of a pH meter, a switched back off voltage is applied to the circuit between the amplifier output and meter. In this way, the meter scale can for example be made to indicate ranges of 1.4 pH anywhere from zero to 14 pH. An open circuit in one of the resistances in the back off circuit, or failure of the back off voltage, results in the meter pointer being hard against one of the meter stops.

A vibrating capacitor converter is a sealed unit and comprises one fixed plate and one movable capacitor plate which is caused to oscillate in a sinusoidal manner by means of an operating coil. This modulates the d.c. potential from the input to an a.c. potential. An example of a modern instrument of the vibrating capacitor type is the E.I.L. Vibret pH meter. The a.c. potential is applied to an electrometer valve which is followed by a transistorized amplifier and rectifier demodulator circuit. The amplifier input stage is screened and requires a high value of insulation resistance as described previously. Vibrating capacitor units have more long-term reliability than electro-mechanical converters but they are more expensive. The electro-mechanical converter is usually driven from the mains supply via a winding on the mains transformer, whereas the operating coils of vibrating capacitors are energized by the output from an oscillator circuit. Failure of this type of modulator to work may thus not be due to a faulty operating coil but to a fault in the oscillator circuit.

pH meter operating difficulties

Common operating faults on pH meters are listed in Table 3.2. In addition to electrode and circuit faults it should be noted that electrostatic effects can be troublesome. Movement of the meter pointer may occur whenever the operator wearing a nylon/silk blouse, shirt, coat, etc. moves near the instrument. This effect can be minimized by using screened electrodes and ensuring that the unscreened portion of the electrode glass is immersed at least 1 inch into the solution. The electrodes may also be covered by an earthed metal screen. Body state electricity may be transferred to a glass electrode by handling. This can be prevented by mounting the pH meter and electrode measuring assembly on an earthed metal plate so that the operator touches it in order to discharge the static before handling the electrodes.

GENERAL ELECTROCHEMICAL APPARATUS

Titration equipment

Potentiometric titrators can be subdivided into several parts, namely, the pH measuring circuit, the titration control unit and the burette/syringe delivery unit. The pH measuring unit is usually of the chopper amplifier type and faults in this circuit will affect the titration. Thyratron valves (gas-filled triode valves having similar characteristics to the silicon controlled rectifier) have in the past widely been used in titrators, an electromagnetic relay in the anode circuit operating a solenoid or a magnetic valve. New instruments becoming available use S.C.R.s but titrators already in use will incorporate thyratrons. The operating principles are however the same. The solenoid valve consists of a metal plunger moving within a coil winding, the movement of the plunger squeezing a flexible tube so stopping the addition of titrant at the required titration end-point. Failure of the thyratron, the supply to it, the relay or the solenoid winding will prevent the titrant from being shut off at the end-point. The grid of the thyratron is connected to the output of a control amplifier circuit which has inputs from the pH measuring circuit and a potentiometer which supplies a voltage representing the required end-point. As the pH value changes and approaches the set end-point, the difference between the two input voltages is reduced, and when they are equal the thyratron conducts operating the shut-off circuit. Failure to stop titration at the end-point can also be due to failure of the control amplifier or no pH input to it. Failure to add titrant may be due to a kinked tube, blockage or to an electronic fault, such as a faulty thyratron operating circuit, causing the shut-off circuit to be energized. The presence of an electronic fault causing the solenoid to be permanently energized can be confirmed by feeling the outside of the solenoid unit which will be hot to the touch. Many titrators use the relay contacts to control the supply to a small-motor-driven syringe which adds the titrant. Microswitches are incorporated as safety limit switches to switch off the motor when the syringe reaches the ends of its travel. If a microswitch contact does not make electrical connection, then the motor will not operate giving rise to failure to add titrant.

Before searching for electronic faults always check for loose, broken and disconnected wires between the titrator electronic unit and the burette or syringe device. Failure to stop the addition of titrant may simply be due to a plug being disconnected. If the solenoid or motor does not operate and voltage is present at the titrators output socket, then check the circuit of the current shut off device or syringe motor for continuity, using a testmeter on the ohms range.

End-point approach proportional control circuits, which reduce the ratio of on to off time of the control relay as the end-point is approached, are often incorporated and the point at which proportional control becomes operative can usually be preset by a potentiometer. Faulty resistors and capacitors can give rise to trouble in end-point approach circuits.

If the required end-point is slightly exceeded this may be due to a fault in the proportional control circuit or to the voltage from the end-point control potentiometer not corresponding to the pH calibration. The end-point potentiometer is supplied from a stabilized power supply and the voltage across the potentiometer track can be checked with a testmeter if the calibration is suspect. Another possible cause is the control knob having slipped round on the potentiometer shaft due to loose fixing screws or excessive force being used.

Carbon-dioxide electrodes

A Severinghaus type pCO_2 electrode consists of calomel and glass electrode elements which measure the pH of a sodium bicarbonate electrolyte solution. The sodium bicarbonate pH changes according to absorption of CO_2 from the sample, which is separated from the measuring glass electrode and electrolyte by a permeable membrane. The relationship between electrolyte pH and pCO_2 is in fact logarithmic and a conversion chart is necessary if an ordinary pH meter is used for measurements.

Frequent calibration using a gas of known pCO_2 is required as the electrode sensitivity depends on the condition of the glass electrode, the membrane and the bicarbonate solution. It is of course necessary to maintain a constant temperature.

If a drift in the meter reading is obtained or varying readings occur with samples equilibrated to the same pCO_2, then the membrane is suspect. To check this, remove the glass electrode from the cell and examine visually. If the defect cannot be seen and the electrode forms part of a gas monitor measuring pO_2 as well as pH and pCO_2, the oxygen measuring circuit can be used to test the membrane. The measuring cell is filled with water from a thermostat-controlled water bath and a test cable is connected

Figure 3.4. Testing the membrane of a pCO_2 electrode.

between the O_2 socket and the pCO_2 electrode assembly, connections being made to the shield of the pCO_2 electrode plug and to the metal body of the cell. A current reading greater than $2\,\mu A$ indicates a hole or leak in the membrane. A membrane in good condition will give a reading of the order of $0.2\,\mu A$.

Another method of testing utilizes a 1½ to 4½ V dry battery. The wall plug and any separate earthing connection from the thermostatically controlled water bath should be disconnected and the cell filled with a liquid containing CO_2. The reading of pCO_2 on the pH meter is then noted. The dry battery is then connected between the metal body of the cell and calomel electrode terminal of the pH meter (the pCO_2 electrode is not disconnected). If a large change in the pH meter reading occurs then the membrane is faulty. This test is shown in Figure 3.4.

Oxygen electrodes

The Clark-type oxygen electrode is a polarographic probe, a current flowing between its anode and cathode dependent on the amount of oxygen contacting the cathode surface. The electrodes and potassium chloride electrolyte solution are separated from the sample solution by a thin membrane stretched over the end of the electrode assembly. The polarizing voltage applied is normally 0.6 V and the small current that flows is passed through a resistance (or variable potentiometer) so that several millivolts are developed for application to a potentiometric recorder or amplifier circuit with meter read-out. Faulty or punctured membranes give a large current in oxygen-free solutions and give rise to a noisy recorder trace resulting from sensitivity to stirring of the sample. With macro-electrodes, a small amount of oxygen is consumed by the probe and since diffusion of oxygen in liquids is too slow to replace that consumed, stirring is necessary. This moves the liquid past the membrane but the rate of stirring should not be excessive or oscillations will be produced in the measuring circuit.

Screened cables should be used in order to avoid interference pick up from associated apparatus such as thermostatically controlled circulating water baths and stirrers. Screening and earthing arrangements are shown in Figure 3.5. Sluggish response can be due to a faulty membrane and before searching for electronic faults in the recorder or amplifer the membrane should be replaced. When fitting a membrane care is necessary (remember it may be less than 0.001 in thick) and no air bubbles should be allowed to become trapped behind the membrane. In order to ensure thorough wetting, a few drops of a wetting agent may be added to the electrolyte.

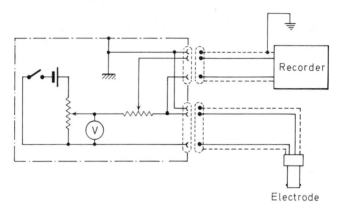

Figure 3.5. Screening and earthing of oxygen electrode apparatus.

Care and use of general electrochemical electrodes

In electro-analytical techniques a variety of electrodes are used and for proper operation of the apparatus, the electrodes must be clean. Platinum and gold electrodes should be stored in distilled water when not in use and cleaned by immersing for several hours in acids (e.g. aqua regia) followed by thoroughly rinsing with distilled water.

Badly contaminated platinum electrodes may require scouring with fine carborundum powder or emery cloth followed by washing in distilled water and cathodic cleaning using a 6 V battery and 5 per cent aqueous sulphuric acid solution. Platinum electrodes, coated with platinum black, used for instance in conductivity measurement, must be kept in distilled water when not in use, as the black coating will flake off if allowed to become dry.

ELECTROCHEMICAL MEASURING AND ANALYTICAL EQUIPMENT

Silver electrodes can be cleaned by using suitable solvents to remove any deposits or by using fine abrasive powder and washing in distilled water. Coated silver electrodes (e.g. silver/silver chloride) should have a clean unbroken surface. If a crack or change in colour is seen, the electrode coating should be removed by using a brush and fine polishing powder, and after washing a fresh coating should be applied by electrolytic deposition.

Tungsten electrodes can be cleaned by scraping or filing.

Clean, pure mercury should be used in dropping-mercury electrodes. Dust and surface scum can be removed by filtering but purifying or cleaning contaminated mercury involves special distillation techniques. The simple polarograph, comprising a dropping-mercury electrode and galvanometer circuit to measure the current, is not usually subject to electrical faults. The difficulties that occur in use are normally chemical in origin, such as the presence of interfering substances and maxima on the polarograph voltage current curve. In subtraction or differential polarography, nodes and antinodes appear on the polarogram if the mercury drops, from the two dropping-mercury electrodes used, do not fall synchronously. To ensure synchronism a striking-off device is necessary. A description of electronic faults on more complex polarographic instruments employing cathode ray tubes and pulse techniques is beyond the scope of this book and servicing is best carried out by a trained engineer. These electronic polarographs which may employ square wave applied voltages are however limited by the instability in the response of the dropping-mercury electrode. In a typical pulse polarograph a 40 msec polarizing voltage pulse is applied 1 s after the mercury drop. The total current is composed of a background diffusion current and a drop capacitance charging current in addition to the faradic current. The background current is subtracted electrically and if the resistance of the polarograph cell is low the capacitive charging current decays quickly. If the current measurement is then made during the last 20 msec of the voltage pulse, only the faradic current is obtained. But it should be remembered that if the cell resistance is high then the charging current is not eliminated. There are also other minor interfering currents. It should be noted that the diffusion current depends partly on the mass of mercury and the pressure head of the dropping-mercury electrode which should be maintained at a constant level. Changes in temperature also affect the diffusion current and polarogram wave height.

Specific ion electrodes used with pH or ion meters are primarily responsive to a particular ion but measure the activity to some degree of more than one type of ion and are not so selective as pH electrodes. The user must keep in mind the error due to the presence of interfering ions. Some substances (e.g. sulphide) in the measuring solution may cause an electrode membrane to be dissolved or poisoned. The sensing membrane element of ion exhange specific ion electrodes has a very limited life and can be replaced when required. Solid state electrodes contain a crystalline salt sensing element and have a relatively long life. The element can be replaced on some types (e.g. Philips) or restored by polishing the external surface with a fine abrasive (e.g. Beckman). Both single and combination glass membrane specific ion electrodes are manufactured and new electrodes may be supplied with the membrane soaking in a teat filled with a 0.1 M chloride solution.

Slow response may occur if deposits have accumulated on a sensing element, if excess interfering ions are present, if the sample solution is not stirred, and if measurements are made going from a concentrated to more dilute solution (memory effect).

Constant-current and constant-potential apparatus

In electro-analytical techniques where a constant potential is required, the potential between the working and reference electrodes in the electrochemical cell is compared with a preset reference voltage in a power unit known as a potentiostat. The basic circuit action is similar to that of the stabilized voltage supply described previously. For constant current operation a resistance is connected in series with the cell and the voltage drop produced compared to the preset reference voltage. If the required operating condition is not maintained and the electrodes are not at fault then the following points should be checked in the potentiostat: (1) the preset reference voltage level, (2) the amplifier circuit and (3) the power control stage.

CHROMATOGRAPHY

Principles of chromatography

In classical column chromatography, which can be regarded as a method of separating substances, a mobile phase carrying the sample is introduced into a large bore, vertical column containing a stationary phase which may be an absorbent solid or a liquid bound to an inert solid supporting medium. In absorption chromatography the substances in the sample are separated by the flow of a suitable solvent. The solvent (eluant) is thus used to introduce the mixture at the top of the column, to separate the components on the column and to elute the separated components from the column.

Apparatus used in a classical liquid column chromatography set-up provides either a gravity feed or constant pumped flow of eluant, a means of detecting the sample constituents by using a photo-electric column monitor, a volume, time or drop controlled device for collecting the eluate from the column in suitable portions or fractions (fraction collector) and a recorder fed from the column monitor which registers absorption peaks.

In chromatography of gases or vapours special temperature-controlled narrow-bore columns (usually coiled as they are 2 or 3 metres long) are necessary. Both solid absorbent and liquid stationary phases are used. The gas emerging from the column is passed through a special detector, the output of which is amplified so that chromatogram peaks are obtained on a recorder. Modern liquid chromatographs also use a narrow-bore column which is temperature controlled and photo-electric or refractometer detectors are used for peak detection.

Column Chromatography

Electronic units

Electronic circuits controlling the speed of peristaltic pumps, photo-electric column drop counters, timers and column monitors are common examples of the application of electronics in liquid chromatography.

Liquid pumps

It is important that pumps used in chromatography and chemical analysis give a consistant flow rate and this in turn means that the pump motor speed must be constant. Various pumping rates are obtained by means of fitting different gear boxes or in the case of a two speed pump by mechanical engagement of additional gears. In pumps utilizing a piston moving in a guide sleeve, maintenance is entirely mechanical and consists of lubricating the piston with a drop of glycerine and the bearings and cam shafts with a light oil.

Erratic operation of the pump is not usually the fault of the driving electric motor but may be due to foreign particles on the valve seats, and leakage from end packing resulting in buffer solutions crystallizing on the piston.

Peristaltic pumps operate by squeezing flexible tubing by means of rollers and in some types a capacitor a.c. single-phase motor is used. The flow rate is adjusted by varying the size of the tubing and the roller pressure. The gearbox attached to the motor has a high ratio and the output speed may be only 1 rev./min. This makes a large amount of torque available and care is necessary when increasing the roller pressure because if the rotor is stalled the gearbox will be damaged.

In order to provide a range of continuously variable flow rates and at the same time make the pump speed less dependent on mains voltage and frequency variation, L.K.B. have introduced an electronically controlled peristaltic pump driven by a stepping motor.

A stepping motor does not rotate continuously but in a number of discrete steps. The permanent magnet rotor aligns itself with the direction of the magnetic field produced at any instant by electrical pulses in the stator windings (Figure 3.6). There are normally two sets of centre tapped stator windings which are fed with square wave pulses from bistable multivibrator circuits (see Chapter 6). Pulses are fed into the bistable multivibrators via a gating circuit from a variable frequency oscillator covering the range 1.1 to 133 Hz.

At slow speeds (up to about 10 per cent of maximum) the pump delivers less than 0.5 μl/step and at high speeds (above 40 per cent of maximum) the pump does not give any pulsations.

The oscillator and motor controlling circuits are mounted on one printed circuit board with the electrical power supply on a separate board.

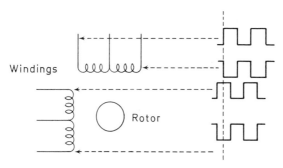

Figure 3.6. Stepping motor waveforms.

If the stepping motor fails to operate and the mains-supply pilot light is on, the fault may lie with the motor windings or connections, the control circuit, the oscillator or the d.c. power supply to the electronics. If only a multirange testmeter is available the resistance of the motor windings can be checked, first disconnecting from the control circuit. It is also simple to check the d.c. output from the power supply. If both these points are in order then it can be assumed that a replacement oscillator and control board is required, although it would be wise to test the potentiometer speed control for an open circuit since this controls the oscillator frequency. If an oscilloscope is available the electronic technician can check the waveforms on the stepping motor windings followed by those at the gating output and oscillator output.

Fraction collectors

Fraction collectors or cutters, which collect liquid drops from a chromatography column into test tubes or special containers, usually incorporate photo-electric drop counting and time-control circuits so that after a preset time or number of drops the collecting tube is changed. In another method the tube is changed by the operation of a mercury switch or a reed switch when a certain volume of liquid has been collected.

A typical photo-detector head incorporates a small low-voltage lamp and miniature photo-conductive cell. When a liquid drop falls between the lamp and photo-detector housing, a shadow is formed across the face of the photo-detector and its resistance changes from a low value to a high value (100 kΩ) as the light beam is obscured. This change in resistance results in an electrical output pulse from the detector circuit as the drop falls. The pulse is amplified by a transistorized amplifier and causes a relay to operate. The relay contacts close every time a liquid drop occurs and are often used to complete the circuit to the coil of an electromagnetic impulse counter.

These counters if of the electrical reset type have two operating coils, one for counting down to zero from the number preset on a thumb-wheel dial, and the other for resetting to the preset number. For proper operation of these counters the operating and reset pulses should not be too short (e.g. 40 msec counting impulse and 180 msec reset impulse) and if a relay is used for resetting, it may be specified that it should have a delay on release of about 150 msec. If for any reason the coils are continuously energized they will be burnt out as they are designed for intermittent operation only.

Electrical contacts are operated when the preset number of impulses has been counted and can be used to operate a delay or slugged relay. This relay resets the counter (Figure 3.7) and also

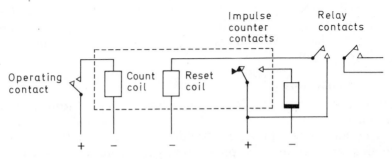

Figure 3.7. Impulse counter operation.

energizes the motor circuit, changing the collecting tube. A common fault with impulse counters is failure to reset correctly and this can be due to faulty electrical contacts on the counter or to the length of the reset pulse. It should be noted that some counter contacts are in fact low-current-rating microswitches and if they are used to switch inductive currents the contacts may stick or weld together. If the reset is not correct, then subsequent operation of the counting coil will not actuate the internal mechanism and the counter appears jammed. This causes liquid drops to be collected in the same tube which eventually overflows and causes a flood.

Microswitches are normally used in the motor control circuit to switch off the motor when the collecting tube has been changed. If the tube does not change or the collecting tube mechanism does not stop after changing a tube, then the most likely cause is faulty microswitch or relay contacts. For preset time operation, impulses are fed to the counter from a microswitch operated by a cam on the shaft of a small combined synchronous motor gearbox unit.

If the counter does not operate on drop counting, first check that the detector head is set up correctly, as correct positioning of the dropping-tube orifice in relation to the light source and photocell is important. If this appears satisfactory then switch to preset time operation and check if the counter operates. If it does, then the fault is located in the drop-counting electronic circuit. The photo-detector output leads should next be disconnected and the detector resistance checked when the light path from the lamp is uninterrupted. The change in resistance should then be observed when the light path is blocked. If there is not an appreciable difference, change the photocell. Should there appear to be a short circuit when checking the photocell resistance, examine the wiring connections and note whether chemicals have been spilt over the detector head. This will give rise to corrosion as shown in Figure 3.8. If the photocell is in order, remove the electronic chassis from its cabinet and, after reconnecting the circuit, note whether the relay contacts feeding the supply to the impulse-counter operating coil, make and break when a large screwdriver (or other solid object) is moved between the lamp and photocell. If the relay operates, but the counter does not, check the

Figure 3.8. Corrosion on photo-electric drop-counter detector head.

operating voltage of the counter. Should the relay not operate then check the voltage across the relay coil. If the voltage does not change when the light beam is interrupted, the fault lies in the amplifier and the transistors should be tested.

Column monitors

Column monitors are electro-optic devices which monitor the transmission of light of the column effluent. They consist basically of a lamp emitting ultraviolet radiation, a filter to select the wavelength of interest and a photocell which gives an output voltage proportional to the light transmission. The photocell output is amplified and fed to a recorder. The components of these monitors are dealt with in more detail in the following chapter which describes electro-optical equipment.

Automatic amino-acid analysers

Chemical analysers are composed of integrated chemical, heating, temperature control, pumping, timing, electro-optical monitoring and electronic recording systems.

There are two types of water baths used, namely a reaction bath and a circulating-water bath, the pump of which circulates water through the glass jackets surrounding the column. The water in the reaction bath is at boiling point and it is important to check that the heating coils are covered. Low water level in the circulating-water bath may cause the thermostat to be above the surface of the water so that the power to the heater is not switched off as the temperature rises. This may cause the water to boil away and air bubbles can be seen in the column jackets. Other results of faulty temperature control or low water levels are poor resolution and high backpressures resulting in broken columns.

Erratic traces on the recorder are often not due to recorder faults, such as dirty or worn channel selector switch or slide wire contacts, but to air bubbles or a foreign body in the cuvette of the electro-optical colorimeter monitoring system. The colorimeter lamps age and blacken with use necessitating different settings on the measuring-circuit baseline or zero-setting multiturn potentiometer. If one of the recorder traces reads full scale, the colorimeter lamp is probably burnt out, although the same symptom is caused by a broken or bad lamp connection. If all the recorder traces are full scale the most likely cause is the failure of the colorimeter lamp's stabilized power supply. On some analysers this is not protected by normal fuses but by thermal links which melt if an overload or fault occurs.

Gas Chromatography

Electronic units

The electronic units of a typical gas chromatograph consist of a detector head, signal amplifier, recorder, analyser oven and temperature programmer.

Flame-ionization gas detectors

As our first example of this type of equipment we will describe servicing and maintenance of a Pye 104 system. Flame ionization detectors (Figure 3.9) are normally used and must be regularly cleaned if blocked jets and contaminated electrodes and insulators

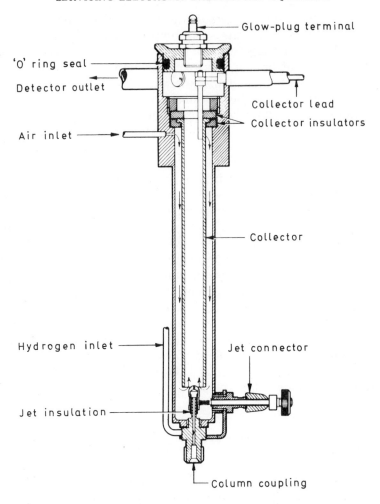

Figure 3.9. Section through a flame ionization detector head (courtesy Pye Unicam Ltd.).

are to be avoided. The detector jet aperture is cleaned with a special tool known as a jet pricker. To remove the insulators and collector electrode, the detector top cap must first be removed. The insulators can then be cleaned with a solvent although an abrasive will be necessary on the electrode. The manual should be referred to for details of the dismantling and reassembly procedures.

The detector flame is lit, after setting the carrier gas and hydrogen flow rates by pressing down the ignite switch on the amplifier while increasing the detector air supply. The switch, which is of the spring-biased off type, should be released when the ignition 'pop' is heard. The gas mixture is ignited by a glow plug on the detector top cap which is energized by a 2.5 V a.c. supply when the ignite switch is depressed. If the glow plug does not operate, check the 10A fuse in the supply circuit. When the column oven is at ambient temperature it is usual to decrease the hydrogen flow rate in order to avoid excessive condensation in the detector. If the flame is reported as going out after lighting, the usual cause is excessive reduction of the hydrogen flow rate.

A stabilized polarizing voltage of 170 V is applied to the detector jet. The current output from the detector is very small (linear range 10^{-14} to 2×10^{-7} A) and thus the electrical insulation must be of a high order. As with electrometer amplifier pH meters the amplifier should not be switched off but left on continuously in order to maintain a dry internal atmosphere.

When only a single column and detector are used, a high standing or background current is obtained which is temperature dependent and thus the sensitivity of the system is limited. If a dual detector unit is used with two identically packed columns the background currents are largely balanced out as the detector outputs are connected in opposition. In practice the resultant standing current (after zeroing the amplifier) is reduced to zero by means of a backing-off circuit. This circuit applies a back-off signal of opposite polarity to the standing current and is derived from a potentiometer fed from a zener diode stabilized supply. If the standing current cannot be eliminated this may be due to failure of the zener diodes, open-circuit potentiometer, faulty contacts or the resistance connected to the selector switch going open circuit.

The ionization amplifier comprises a subminiature electrometer valve input stage followed by a high-gain transistorized d.c. amplifier. Servicing of this unit should be left to a trained technician. The amplifier power is obtained from a +30, 0–20 V stabilized supply. The amplifier gives 1–10 mV outputs to a potentiometric recorder.

Column oven-temperature control

The analyser oven contains a 2000 VA heating element which surrounds an air circulation fan. A thermal fuse which melts at 420°C (or alternatively at 570°C) provides protection against overheating. The only maintenance required is regular oiling of the fan motor bearings. The oven contains a platinum resistance thermometer which connects to the temperature controller. A simplified diagram of a temperature programming and control circuit is shown in Figure 3.10. The resistance thermometer forms

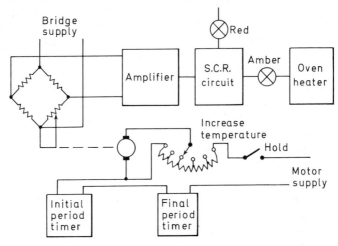

Figure 3.10. Gas chromatograph column temperature programming and control circuit.

one arm of a measuring bridge circuit with the oven temperature potentiometer in another arm. The output from this bridge circuit is amplified and operates a silicon-controlled rectifier which controls the supply of power to the heater.

Two motorized timer units with electrical contacts which operate after preset 'initial' and 'final' periods are provided to give isothermal operation at the initial and final temperatures of the temperature programmer. After the initial heating period, the first timer cam operates a switch which energizes the motor driving the oven temperature potentiometer, at a rate determined by the increase temperature switch. This switch selects resistances in the motor circuit in order to control its speed. When the required final temperature is reached the motor is switched off and the final period timer started. Recent temperature programmers utilize electronic timers and digital techniques to give more accurate and reproducible results.

A switch is incorporated in the circuit to override the program at any time and hold the temperature constant. Two neon lamps are provided on the analyser oven control unit, a red lamp and an amber 'heater power' lamp. If the amber lamp does not come on this may be due to the program controller and analyser oven being connected to different phase electrical supplies resulting in incorrect operation of the S.C.R. circuit. There may of course be a fault in the S.C.R. circuit such as failure to trigger, and this part of the circuit should be examined for short and open circuits and loose connections. If the amber light does not go dim on isothermal operation, as it normally should when power to the heater is reduced, the resistance thermometer and associated wiring should be checked for a short circuit. The resistance thermometer electrical resistance depends on the oven temperature but should normally exceed $100 \, \Omega$. There may also be a fault in the S.C.R. trigger circuit or wiring to the oven temperature potentiometer.

When operating for programmed temperature increase, failure of the increase temperature period to start after the initial period, is due to a faulty timer, microswitch or motor circuit.

If the initial timer fails to start, the timer or hold/start switch may be faulty. Should the final-period timer not switch the oven heater off, then the cam operated microswitch is suspect.

A 12.6 V low wattage injection heater is also used, enabling the sample injection point to be at a higher temperature than the rest of the column. The heating element is controlled by an energy regulator which regulates the supply to the heater step-down transformer. Two thermocouples are provided for measuring the temperature, and one is positioned at the injection point while the reference junction is suspended within the column oven.

There are of course mechanical parts such as sealing rings and discs which require replacement or cleaning on routine maintenance and for details of these the instrument manual should be consulted. Due to the extreme sensitivity of the flame ionization detector it is particularly important to ensure that there is no contamination in the system. The presence of only a few parts per million of any organic substance will cause trouble. In any make of gas chromatograph employing this type of detector, unsatisfactory performance is often due to such contamination and not to instrument faults. Contamination may be the cause of a high background current and a high noise level. The most likely places for contamination to occur are in the air supply, the tubing of the air and hydrogen supply, the chromatograph columns and the burner unit of the flame ionization detector. A contaminated burner unit often gives a variable wandering on the recorder. One method of cleaning out a column is shown in Figure 3.11. A fault-finding guide for a flame ionization system is given in Table 3.3.

Electron-capture detector

Another type of detector used with an electrometer amplifier is the electron capture type (Figure 3.12) and similar remarks to those above regarding contamination, cleanliness and maintaining the insulation resistance of the detector insulation elements apply.

Erratic spikes produced on the recorder trace following a period of normal operation may be due to the spark discharge moving around on the electrodes. This can usually be cured by setting the source current to a higher value than normal for a short period (several minutes). If the discharge does not stabilize, the system may require purging or there may be leakage of oxygen (air) into the flow system. This can be checked by subjecting each suspect joint or fitting to a brief jet of nitrogen gas. The leakage point then gives a change in the background on the recorder. Stabilized bias and polarizing, and source-current power supplies are used with this type of detector, in addition to a starting circuit which applies a high voltage to discharge electrodes in order to initiate the discharge. Failure to start discharge may be due to a fault in the electronic starting circuit or to contamination in the system as

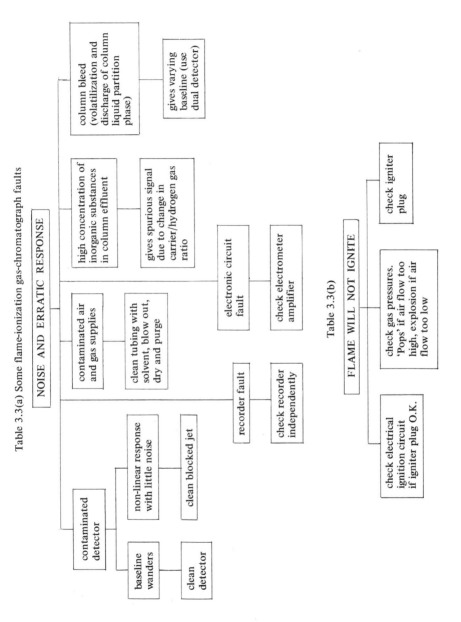

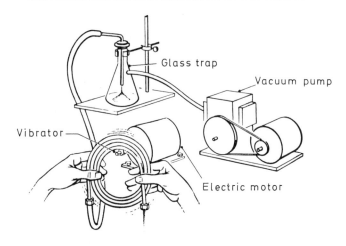

Figure 3.11. One method of cleaning a gas chromatograph column (courtesy of Beckman Instruments).

the detector firing voltage level is raised if there is a high concentration of impurities in the system.

Thermal-conductivity detector

Finally some mention should be made of the thermal conductivity detector. A detector of this type, used for example in the Beckman GC2A gas chromatograph, is shown in Figure 3.13. This detector uses four rhodium-plated tungsten filaments connected in a Wheatstone bridge circuit. The filaments are mounted in separate chambers in the detector block and the sample carrier-gas mixture flows through one pair of chambers and the pure carrier-gas through the other pair. The bridge circuit is supplied from a stabilized voltage supply and the rate of heat dissipation from the filament depends on the thermal conductivity of the surrounding gas. As the bridge is initially balanced with carrier gas flowing over both pairs of filaments, the presence of the sample on one side of the detector causes a change in thermal conductivity and hence a change in the filament temperature and resistance. The resulting output signal is passed through a switched resistance network (attenuator) which reduces the signal voltage applied to a recorder.

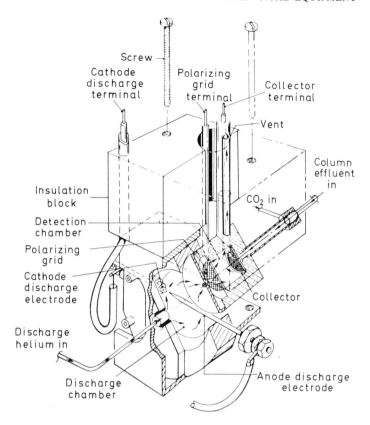

Figure 3.12. Operation of the electron capture detector (courtesy of Beckman Instruments).

The column temperature is controlled by an electronic circuit using a resistance thermometer. The supply to the detector bridge is highly stabilized, a portion of the bridge supply being compared with a reference voltage obtained from mercury batteries. Any difference is applied to an amplifier, the output of which controls the bridge voltage and, hence, filament current.

A noisy trace on the recorder may be due to a recorder fault, or to contamination of the detector block. The latter can originate from running a column at too high a temperature giving rise to

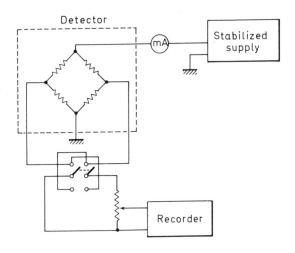

Figure 3.13. Thermal-conductivity detector circuit.

column 'bleeding' or to decomposed material having been deposited on the inside of the detector cell. If the detector has to be dismantled for cleaning, great care should be exercised when removing the wire filaments. They should not be touched by hand or the arrangement of the wire coils disturbed. Filaments are supplied in matched sets of four and it is sometimes recommended that the whole detector should be changed rather than attempting to replace filaments.

Usually the current output from the bridge is measured by a milliammeter and some effects of faults on the meter reading are shown in Table 3.4.

Interpretation of gas chromatograms

Incorrect operating conditions will cause unusually shaped chromatograms and it is important that the technician can determine from the chromatograms the correct action to take. For instance, if sharp unresolved peaks occur shortly after injection, the column temperature is probably too high. The same symptoms may occur of course if there is an electrical/electronic fault causing overheating. A long-drawn-out tail on the trailing edge of

Table 3.4. Effects of faults on thermal conductivity detector circuit current

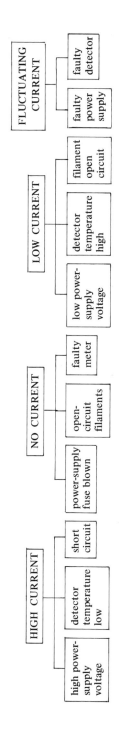

the peaks is caused by using the wrong stationary phase. The stationary phase should also be changed if good peaks are not obtained and the resolution is poor when the analysis and retention times are normal. If peaks are obtained but the resolution is poor, due to lack of separation between the peaks, then the column temperature should be lowered or the gas flow rate changed. This situation may also be improved by using a longer column or less sample material. Too much sample material can cause asymmetric peaks (one side sloping and the other side almost vertical) although they are often caused by thermal decomposition of the sample, in which case the column temperature should be reduced. When a combination of very small and large component peaks is obtained, the size of the smaller peaks can be increased by increasing the column temperature and gas flow. The same result can be achieved by increasing the amount of sample material, but care is necessary to avoid overloading the column. Wandering baselines are usually due to contamination or failure to maintain the column temperature constant. These points should be checked before deciding that the chromatograph amplifier or recorder is drifting. Adequate time must be allowed for the column to stabilize before starting analysis. Noisy traces may be due to contamination or to electronic faults and the cleanliness of the detector should always be checked first.

4 Electro-Optical Measuring Instruments

The electro-optical instruments used in scientific laboratories to measure the transmission of light through liquid samples, the effluent of chromatography columns, density of photographic film and plates, etc., basically comprise a light source, optical system and photo-electric detector circuit. The light wavelength at which measurements are made is determined by an optical filter, prism or diffraction grating.

In chemical and biochemical applications, ultraviolet–visible spectrophotometers are used in the quantitative determination of organic and inorganic substances, infra-red spectrophotometers for the identification and investigation of the structure of complex molecules, flame and atomic absorption spectrophotometers for the determination of specific elements and densitometers for scanning photographic films obtained from analytical instruments (such as emission spectrographs and analytical ultracentrifuges).

INSTRUMENT COMPONENTS

Tungsten filament lamps

The most common light source for work in the visible region of the spectrum is the tungsten filament lamp. In simple instruments such as a filter colorimeter, the electrical supply to the lamp is stabilized by a constant voltage transformer. More sensitive instruments however require a greater degree of stability and a constant voltage electronic regulating circuit of the type previously described is used.

Figure 4.1. Blackened tungsten lamp.

Tungsten filament lamps (typically car headlamp 36 W type) have a limited life, as tungsten evaporates from the filament and is deposited on the inside of the lamp bulb (Figure 4.1). This together with pitting of the filament may be the cause of erratic readings on the instrument. When used with a 6 volt electronic voltage stabilized unit, a 7 amp 'slow blow' fuse is generally used to protect the power supply. As the lamp ages, its electrical resistance changes and this may be the cause of the fuse blowing repeatedly. Another cause of instability in the lamp circuit is worn electrical contacts on the tungsten lamp switch. When replacing tungsten lamps, wipe off any finger marks as they will subsequently burn into the glass when the lamp is switched on.

Quartz halogen lamps are also used in spectrophotometers. They give more intense light output than the tungsten filament lamp and are not subject to blackening of the bulb.

U.V. lamps

In the ultraviolet region of the spectrum, deuterium or hydrogen arc discharge lamps are used. A typical lamp is shown in

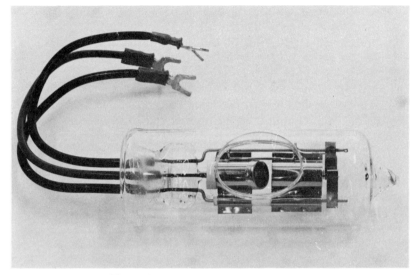

Figure 4.2. Deuterium discharge lamp.

Figure 4.2. These lamps require a special power supply which gives a voltage of the order of 400 V for striking the lamp, a filament supply and regulation of the arc current. On some power supplies a variable switched filament supply is provided so that maximum voltage is applied for starting and reduced voltage for operation in order to increase the maximum life of the lamp. The voltage should not be reduced to too low a value or flickering may occur and bright incandescent spots appear on the filament.

The life of a u.v. lamp is limited and may only be about one tenth of the life of a tungsten lamp. These lamps should not therefore be left on if they are not required for an appreciable length of time, particularly as they are expensive. When a hydrogen or deuterium lamp is nearing the end of its useful life its light output decreases sharply and it may fail to strike. Failure of a lamp to strike correctly may occur if it is still very hot from previous use. This is not a fault, and after allowing the lamp to cool it should strike normally. When attempting to strike a hot lamp, ionization often occurs inside the anode structure instead of in one spot in front of the lamp quartz window.

A simplified diagram of a Beckman constant-current hydrogen lamp supply is shown in Figure 4.3. The zener diode provides an 11 V reference at the base of the first transistor in a series chain. The circuit resistance comprise a calibration resistance made up of lengths of resistance wire cut to an exact length and connected to give a lamp current of 300 mA, plus series-connected wire-wound high-wattage resistances. If the lamp current tends to rise, then the transistor chain conducts and shunts the high-wattage resistors, through which the lamp current is passing, so regulating it to 300 mA.

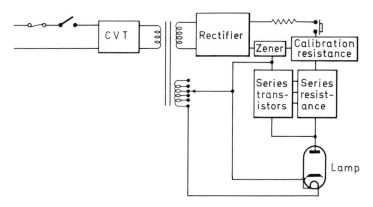

Figure 4.3. Simplified diagram of Beckman deuterium lamp power supply.

If a lamp does not fire, first ascertain that sufficient warm up time has been allowed. Check also that current is flowing through the filament by observing the small red glow when the maximum voltage is applied. If no glow can be seen check that low voltage a.c. (usually 3 V max) is present at the connection block (or terminals) in the lamp housing. If the filament voltage is correct, then connect the test voltmeter between the lamp anode and cathode connections and check that the voltage rises to nearly 400 V d.c. when the start button is pressed. If it does then the lamp is at fault. If of course a lamp that has been in use a considerable time fails then it should be changed without making the above measurements. Should the stability of the power supply and/or lamp be suspect, remove the lamp and connect in its place a 0-1 A ammeter in series with a 300 Ω 50 W resistance. Variations in current can then be observed and the circuit regulating action checked by supplying the power unit from a variable transformer (Variac). A 20 volt change of supply voltage should typically only give 1 per cent or 0.1 per cent change in current depending on the design specification.

Special lamps

A deuterium lamp can be used down to a wavelength of 185 nm and has peak emission at certain wavelengths.

Two useful points for checking purposes are the emission lines at 656.3 nm and 486.1 nm. Mercury discharge lamps also emit a number of discrete lines and in photometers a particular wavelength is isolated by using filters. Xenon discharge lamps used in fluorimetry and polarimetry emit high-intensity radiation and should be handled with care as they can explode if knocked or dropped, due to high gas pressure inside the glass envelope. The intense radiation from xenon lamps, ionizes the oxygen in the surrounding air to produce ozone and in confined spaces adequate ventilation is essential if a build-up of dangerous ozone is to be avoided. Discharge lamps in which the arc is struck between two electrodes, may become unstable in use due to the discharge wandering in relation to the electrodes.

When aligning high-intensity u.v. lamps, particularly those emitting radiation between 313 and 334 nm, protective tinted glasses with side shields should be worn. If this is not done dangerous conjunctivitis of the eyes can occur.

Infra-red light sources and detectors

Some types of infra-red source emit radiation when a current is passed through them at normal ambient temperature. Examples are the globar rod and nichrome wire elements. An open circuit or failure of the source supply are typical faults. The Nernst glower is another common infra-red source and requires heating to a high temperature. The zirconium rod in this type of source unit is normally non-conducting but conducts current when heated to about $800°C$. Failure to emit may be due to open-circuit heater elements or a fault in the heater control circuit. Thermistors, bolometers, photocells and special Golay detectors are also used to detect infra-red radiation.

Photocells

Common vacuum-tube photocells comprise a photosensitive cathode and a normal positive-anode electrode. There is a linear relationship between the intensity of the incident light and the anode current. They normally give little trouble but in time their sensitivity falls off and difficulty may be experienced in balancing null-circuit spectrophotometer instruments.

The barrier layer photo-emissive cell (usually selenium deposited on iron) develops an electrical output when exposed to light. With this type of cell, less current than normal may be obtained after exposure to strong light and this fatigue effect can be troublesome. The cell should be left in darkness to recover. Other difficulties may arise due to its high temperature coefficient, and in flame photometers special mounting arrangements are necessary to keep the cell cool.

Photoconductive cells

Photoconductive cells which change their resistance when exposed to light are also used as detectors. An example is the lead sulphide cell which is used for instance on the Beckman DK2 spectrophotometer described later. A polarizing voltage is applied across this cell. On this instrument if the lead sulphide cell deteriorates and its resistance falls below 150 kΩ, poor resolution is obtained.

Photomultipliers

Photomultipliers are used in very sensitive instruments and consist basically of a photosensitive cathode, a 'dynode' electron multiplying system and an anode. An electron emitted from the cathode is directed to the first dynode electrode where secondary electrons are produced. These secondary electrons are in turn directed to the next dynode and in this manner considerable electron multiplication is obtained. The final electrons are collected by the anode. The circuit connections to a photomultiplier are shown in Figure 4.4.

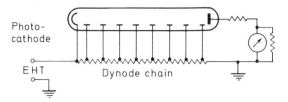

Figure 4.4. Simple photomultiplier circuit.

A high-voltage (EHT) supply is necessary and is normally stabilized. Since photomultipliers are very sensitive devices they must not be exposed to strong light with the EHT voltage applied, or the photosensitive cathode may be stripped. On spectrophotometers this is often prevented by using a microswitch in an interlock circuit which cuts off the EHT supply when the cell compartment cover is raised. Failure to obtain an output from an instrument of this type may be simply due to a faulty microswitch and not to failure of the photomultiplier or EHT supply. Incorrect operation will result if one of the dynode-connected chain of resistances goes open circuit. Another source of trouble may be shorting or tracking of the high voltage between components wired to the photomultiplier socket since they are very close together. This occurs if dust or dirt accumulates over the components or when there is high humidity. Electrical noise on the photomultiplier output may be due to a bad connection between the socket and pins on the tube base.

In both vacuum phototubes and photomultipliers a small current flows in the output circuit when the device is not exposed to light (i.e. in a dark condition). This current is very dependent on temperature and must be backed off within the measuring instrument before measurements can be made.

Mirrors and prisms

Erratic wandering behaviour or a rapid fall in the level of the trace of a recorder, operated from an electro-optical instrument, may not be due to an electronic fault but to dirty or contaminated mirrors in the optical system. Dirty mirrors cause the incident light to be scattered and contamination of a mirror may result in the incident light being absorbed by, for example, a chemical spilt on its surface. In general the mirror surfaces should never be touched and dust should be blown off with an air syringe. Mirrors external to a monochromator (e.g. in the lamp housing or cell compartment) may be cleaned *in situ* by washing the front surface with a distilled water-teepol mixture applied by a fine camel hair brush. This is followed by rinsing with a fine jet of distilled water from a plastic wash bottle. Excess liquid is soaked up by the use of absorbent paper tissues. If the mirror is clean all the distilled water should run off without leaving droplets on the surface. Very dirty mirrors can also be cleaned by the use of collodion solution which is applied by a soft brush. Several coats are necessary and when dry the collodion layer is peeled off removing dirt and dust which adheres to the collodion. It is also sometimes possible to remove fingermarks by this method. Contaminated mirrors may be cleaned by wiping with a cotton wool swab moistened with anhydrous diethyl ether or methanol, although often replacement of the mirror is required. In some instruments the reflectance of the mirrors is matched to close limits and the mirrors may have to be changed in a matched set. On no account should any mirror surfaces be touched by hand. Replacement of a mirror will necessitate re-alignment of the optical system and this should only be undertaken by competent trained persons or the manufacturer's service engineer. It is most important that users do not interfere with the positioning of optical system components within the monochromator. They are accurately located and in the setting-up procedure the fixing screws are tightened and locked with paint or varnish.

On infra-red instruments the optical components (e.g. rock salt prisms) are hygroscopic and it is important to keep the monochromator dry. This may be done by heaters within the monochromator and it is therefore important not to switch off the instrument at the wall plug when it is not in use.

ELECTRO-OPTICAL INSTRUMENTS

The colorimeter

A simple colorimeter is shown in Figure 4.5. The component parts of this type of instrument are a tungsten lamp, optical filter,

ELECTRO-OPTICAL MEASURING INSTRUMENTS

Figure 4.5. Photo-electric colorimeter.

constant-voltage transformer lamp supply and barrier layer photocell connected to a moving-coil meter (calibrated in transmission and optical density). The two barrier layer cell connections are shown on the left of the photograph and the constant-voltage transformer on the right. Generally erratic readings are due to the lamp or bad electrical connections (not forgetting switch contacts) while lack of sensitivity may be the fault of the photocell or lamp.

If the sampling technique of the operator is incorrect then inaccuracy in the readings on the meter will be obtained. With instruments taking test tubes, only optical-grade tubes should be used, which must be clean and dry externally. There should be no air bubbles in the liquid sample. It is of course important that the correct filter is used for the test being made.

Chromatograph column monitors

The effluent of the column passes through a flow cell and is monitored by a detector unit comprising a mercury lamp, optical filter and vacuum phototube. The output from the phototube is amplified and typically applied to a current-operated moving-coil recorder. Failure of the lamp is the commonest cause of lack of output although the connecting leads between the column-mounted detector and the separate amplifier control unit should also be checked. Air bubbles in the flow cell can also be troublesome causing erratic readings.

Manual null-balance spectrophotometers

Examples of this type of instrument are the Unicam SP600 and SP500 spectrophotometers. The SP600 normally utilizes a tungsten or quartz iodine lamp and the light wavelength required is obtained by rotating a prism in the monochromater. The output current from either a red sensitive or blue sensitive photocell is passed through a high-ohmic-value resistance connected to the input of an electrometer valve. A simplified circuit is shown in Figure 4.6. The dark current is backed off to zero the null meter with the control switch in position 1, when a shutter blocks off

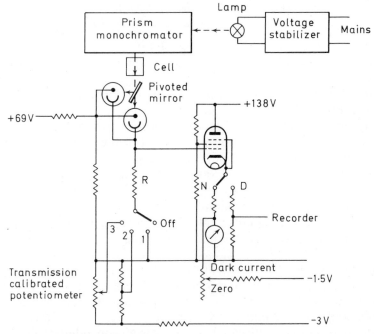

Figure 4.6. Manual null-balance spectrophotometer circuit (based on Unicam SP 600).

the light beam. In order to set the instrument for 100 per cent transmission of light, the control switch is switched to position 2, and with a 'blank' cell in the light path the monochromator slits are adjusted to zero the meter. In this switch position, fixed resistances are in circuit equivalent to the 100 per cent position of the transmission calibrated potentiometer. Measurements are made with the sample in the light path and the control switch in position 3. The transmission potentiometer is then adjusted to zero the meter and the amount of light transmitted through the sample is read off from the calibrated scale.

If the meter cannot be balanced in use and the pointer is hard over to the left then too much light is falling on the photocell. This is caused by incorrect operation of the optical system slits or their operating mechanism. If insufficient light falls on the photocell, the meter pointer is hard over in the opposite direction and indicates either failure of the lamp or that the dark current shutter is obscuring the light beam in the measuring position. If the meter can only be zeroed with very wide slit widths, it is possible that the photocell has lost sensitivity. Faults in the amplifier or HT supply can also give the symptoms above. No response at all in the meter may be due to failure of the lamp, photocell, amplifier, HT supply or to the meter itself. Construction of the Unicam SP600 is shown in Figure 4.7.

The Unicam SP500 is a more sensitive null-balance instrument incorporating a deuterium lamp to extend the wavelength range (186-1000 nm). The series 1 version used a valve amplifier, while the newer series 2 instrument uses transistorized electronics.

A fault-finding guide applicable to the series 1 Unicam SP500 is given in Table 4.1. Always check that the correct lamp and photocell have been selected for the wavelength used, before searching for instrument faults.

It is most important to change regularly the silica gel desiccants located in the amplifier compartment and monochromator. Failure to do this will allow moisture to affect the operation of the sensitive electrometer measuring circuit. When the crystals are pink, they are saturated with moisture and they should be replaced by fresh dry blue crystals. The old pink crystals can be dried in an oven.

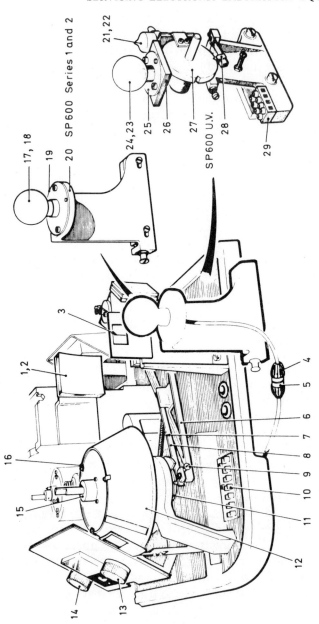

Figure 4.7. Construction of Unicam SP 600 spectrophotometer.

1., 2. concave mirror and clip
3. cursor
4. lamp plug
5. lamp socket
6. actuating arm
7. spring
8. adjustment spring assembly
9. adjusting screw
10. terminal strip
11. cam and bush assembly
12. wavelength scale
13. variable resistance control
14. switch
15. cone bush and spindle
16. scale clip
17. tungsten lamp assembly
18. tungsten lamp
19. lamp ring
20. lamp bracket
21., 22. deuterium lamp and clip
23. lamp assembly
24. tungsten lamp
25. lamp ring
26. lamp bracket
27. control shaft assembly
28. click spring assembly
29. terminal block

(courtesy of Pye Unicam Ltd.).

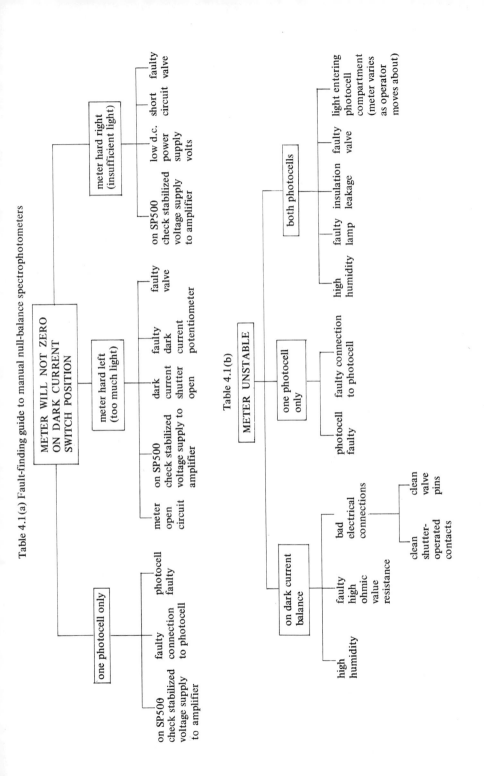

SERVICING ELECTRONIC LABORATORY EQUIPMENT

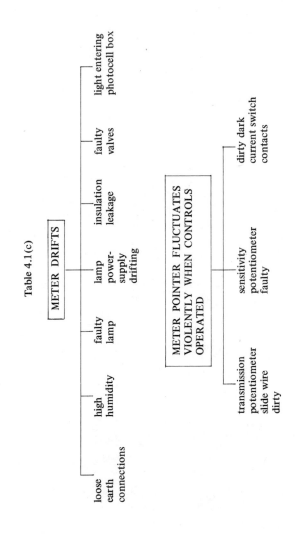

Table 4.1(c)

ELECTRO-OPTICAL MEASURING INSTRUMENTS

Table 4.1(d)

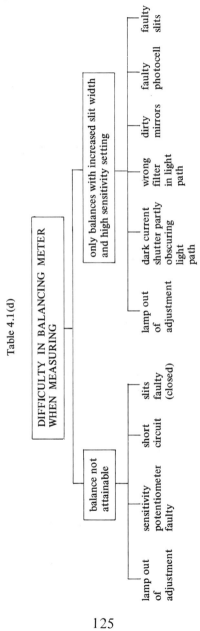

Figure 4.8. Dirty and corroded contacts on a rotary control switch.

The most difficult fault to cure on this type of instrument is instability. If unstable behaviour of the meter only occurs when the dark current shutter is closed, it may safely be assumed that the fault is not due to the lamp or its power supply. Bad electrical contact is often the cause, and the potentiometer tracks can be cleaned with methylated spirit. Control switches and contacts operated by the dark current shutter can be cleaned by mechanical operation after applying a drop of solvent on the end of the screw driver. Dirty and corroded contacts on a control switch are shown in Figure 4.8. Switch cleaning fluids can be used but should be applied with caution. Excess application should be avoided and aerosol spray cans are not recommended in this instance. The high ohmic value resistances (typically 2000 megohms) and insulators should not be fingered and should be cleaned using tissues. If instability is due to the silica gel crystals being saturated with moisture and thus unable to keep the amplifier compartment dry, it must be appreciated that it will be at least an hour before fresh crystals will dry out the compartment sufficiently. Faulty and

Figure 4.9. Replacing the electrometer valve in a Unicam SP 500. 1 spectrophotometer.

erratic operations can also originate in the electrometer valves and in Figure 4.9 a valve is being replaced in a series 1 Unicam SP500 spectrophotometer. Correct alignment of light sources is essential on every type of spectrophotometer. Tungsten lamps are usually positively located in the vertical plane but some adjustment may be necessary to the lamp house mirror in the horizontal plane. Corrent alignment and focusing are checked by observing the image obtained on a white card and placed in front of the shutter. A clear-edged, bright, uniformly-illuminated image should be obtained. If it is not, then the lamp house mirrors must be adjusted. The image obtained with a misaligned lamp is shown in Figure 4.10.

Hydrogen and deuterium lamps often slide in the vertical planes as they are held by spring clips. They also twist radially in the clips and after coarse positioning, final alignment can be performed by

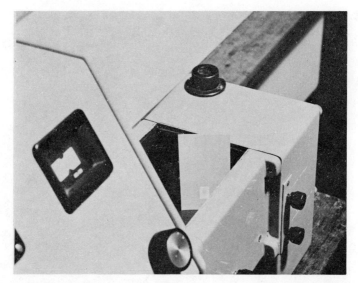

Figure 4.10. Checking lamp alignment on an SP 500.1 spectrophotometer.

screw adjustments. The image obtained in the ultraviolet region of the spectrum will not be as bright as that in the visible region. Remember to clean off fingermarks with a grease solvent before switching on a replacement lamp.

While the operating principle of the series 2 instrument is similar to that of the series 1, the electronic circuit and therefore servicing is different.

The changeover from one photocell to the other is performed by enclosed reed switches which are operated by a magnet on the wavelength-adjusting mechanism. Thus there are no exposed contacts to cause trouble. It should be noted that the blue sensitive photocell is always connected and when it is in use the output of the red photocell is earthed. On selection of the red photocell the photocells are connected in parallel and the earth connection is broken. A microswitch is operated by the shutter and shorts out the transmission/absorbance potentiometer when setting the zero. There are two modes of operation in this instrument, direct readout and null balance, the latter being the most sensitive. The zero potentiometer on direct readout sets the 0-100 meter to 0 per cent transmission and on null balance the centre-zero meter is set to mid-scale. The zero potentiometer in fact feeds a back-off voltage into the base of the amplifier output transistor to cancel its input due to the photocell dark current. Failure to set zero can therefore be due to an open-circuit zero potentiometer, failure of the zener stabilized supply across the potentiometer or a faulty transistor. The appropriate lamps are switched on separately and a solenoid is used to operate the lamp change mirror (Figure 4.11). In the centre position of the lamp change switch, automatic changeover occurs by the action of a microswitch actuated at 335-340 nm by the wavelength control.

If the lamphouse mirror remains permanently in the deuterium lamp position then the fault is probably in the solenoid. To check this, use a d.c. voltmeter across the connections to the solenoid. A reading should be obtained with the tungsten lamp switch on. If no voltage reading is obtained, check the lamp change switch and full wave rectifier supplying the circuit.

A number of components are included in the amplifier circuit to maintain stable operation. These are, a capacitor between the base and collector of the BFY50 transistor amplifier of the power supply section, and a capacitor between the screen and filament of the electrometer input valve. If drifting occurs when the ambient temperature changes, then this may be due to failure of the special temperature coefficient resistor (silistor) in the emitter circuit of the amplifier second stage.

Split (double-beam) spectrophotometers

In this type of instrument, the light beam is alternated between a sample and reference cell by an optical switching system. Thus in one position of the system, the photosensitive detector is subjected to light that has passed through the blank or reference cell, and in the other position to light that has passed through the sample.

In a ratio-recording instrument (Figure 4.12), when the sample and reference beams are unbalanced, an error signal is produced which is amplified and applied to a servo-amplifier recorder. The recorder pen is coupled to the wiper of a transmission or

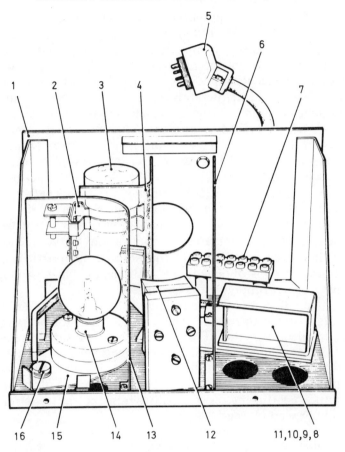

Figure 4.11. Lamp house on Unicam SP 500.2 spectrophotometer

1. lamp house assembly
2. clip
3. deuterium lamp
4. screen
5. plug
6. solenoid screen
7. terminal block
8. solenoid assembly
9. solenoid
10. spring
11. flat mirror
12. spherically concave mirror
13. screen
14. tungsten lamp
15. lamp holder assembly
16. adjustment pivot screw

(courtesy of Pye Unicam Ltd.).

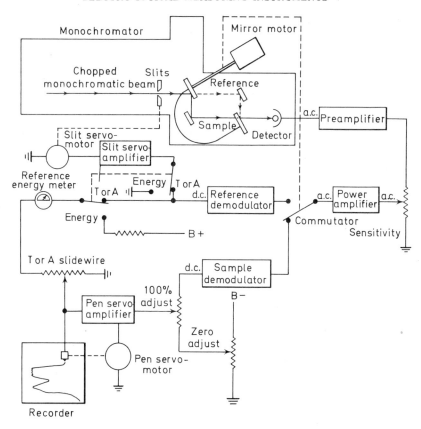

Figure 4.12. Beckman DK2 ratio recording spectrophotometer (courtesy Beckman Instruments).

absorbance calibrated slide-wire potentiometer which then moves to rebalance the system by reducing the recorder electrical input to zero. With null balance instruments the error signal is reduced to zero by the servosystem motor driving a wedge or comb so as to balance optically the intensity of the sample and reference light beams. A spectrophotometer using this method is shown in Figure 4.13. In both methods the optical beam chopping or switching mechanisms are coupled to a synchronous rectifier or

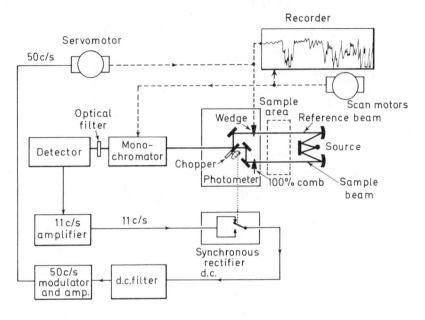

Figure 4.13. Perkin Elmer 257 Optical null spectrophotometer (courtesy of Perkin Elmer Ltd.).

demodulator in the electronic circuit. The operation of the Beckman DK2 ratio-recording spectrophotometer is as follows.

After amplification, the photodetector signal is passed through an electrical commutator switching and demodulator system (synchronized with the optical switching system) to either the reference or the signal amplifier. A recorder measures the difference between the demodulator amplifier outputs and the pen is deflected by an amount proportional to the transmission of light through the sample. Automatic operation is achieved as a small motor drives the monochromator prism between selected wavelengths. Adjustment of the monochromator slits is also automatic as the output from the reference channel amplifier is compared with a stabilized voltage. Any difference is amplified and used to control the slits by means of a servo-motor so that the reference beam energy is maintained constant.

ELECTRO-OPTICAL MEASURING INSTRUMENTS

Table 4.2. Typical faults on the ratio recording Beckman DK2 spectrophotometer

Noisy Recorder Trace

Amplifier valves, transistors or components noisy (particularly pre-amplifier)
Light source lamp(s) noisy
Photo-detector noisy
Synchronous rectifier or commutator switches faulty (particularly electrical contacts)
Slide wires dirty
Pen servo-amplifier faulty (check gain and damping settings)
Slit servosystem oscillating (gain too high)
Noisy d.c. supply to electronic circuits (particularly neon stabilizers)
External interference

Cannot Set Zero Transmission

Absorbance and transmission slide wires misaligned
Zero potentiometer faulty (check resistance and wiper operation)
Stabilized supply to zero potentiometer faulty (check voltage)

100 per cent Transmission Line Not Straight and Level

(a) If 100 per cent position changes with time at fixed wavelength
Rotating mirror bearings and alignment faulty
Filter inductance coils faulty
Fault in time constant circuit components

(b) If 100 per cent position stable with time at fixed wavelength but changes when scanning
Contaminated sample compartment mirrors (100 per cent line wavelength dependent)
Slow response of slit servo (check gain setting)
Difference in sample and reference time constant circuits (100 per cent line deviation in region where water vapour absorbs)

Poor Resolution

Bad lamp condition, alignment or focusing
Photodetector or it's supply faulty
Low reference energy (check amplifiers and d.c. supply)

With this type of instrument typical difficulties experienced are noisy and uneven 100 per cent transmission line traces on the recorder. The possible causes of these and other malfunctions are listed in Table 4.2. Always check first that the correct lamp and detector have been selected and that the normal time constant, sensitivity and range settings are being used. On the DK2 spectrophotometer the electro-mechanical optical-beam switching arrangement (Figure 4.14) requires regular maintenance. The flexible shaft requires periodic replacement when the amount of twist or free play between the two ends becomes excessive. This gives rise to slow variations in the recorder trace. Regular

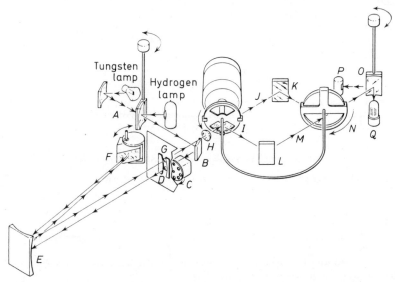

Figure 4.14. Optical path and beam switching arrangement on Beckman DK2 spectrophotometer *A*. source condensing mirror *B*. entrance mirror *C*. entrance beam chopper *D*. entrance slit *E*. collimating mirror *F*. quartz prism *G*. exit slit *H*. exit beam condensing lens *I*. rotating mirror *J*. reference beam *K*. reference stationary mirror *L*. sample stationary mirror *M*. sample beam *N*. rotating mirror *O*. detector selector mirror *P*. lead sulphide cell *Q* photomultiplier tube (courtesy of Beckman Instruments).

adjustment of the cam-operated demodulator or commutator switches is necessary and this must be performed in conjunction with an oscilloscope. The oscilloscope is connected to the reference channel when the left-hand switch is adjusted and to the sample channel when adjusting the right-hand switch. The correct waveform should have twelve to thirteen peaks and must end on a downward sweep. The fixed and rotating mirrors must be clean and should be matched to within 1 per cent. The alignment of the rotating mirrors with each other and the rest of the optical system is also important.

The reference voltage for the slit servo is obtained from a battery inside the electronic unit and the condition of the battery should be checked periodically. The response of the slit servo-system is set by an internal screwdriver potentiometer adjustment. The reference energy level indicated on a meter is also set up by a screwdriver adjustment.

The light beam from the monochromator is chopped at 400 Hz (on 50 Hz supply instruments) to obtain a good signal to noise ratio and to eliminate the effect of stray light. Stray light reflected from optical compartment walls is not chopped and therefore gives a direct voltage signal which is not passed through the amplifiers since inductors tuned to 400 Hz are incorporated. Replacement of the inductance coils should only be undertaken by a service engineer as it may be necessary to try several coils in order to keep the instrument within specification. This is due to the impedance of the coils changing at different rates as the input voltage changes.

Due to the beam switching there is a 15 Hz component in the direct voltage from the demodulator circuit outputs and this is removed by filtering. A faulty filter unit generally results in a low output from the relevant channel.

Noise is a problem in all sensitive spectrophotometers and often originates in the detector or the first stage of the amplifier. If the input to the amplifier is short-circuited and the noise persists on the recorder trace, then the detector circuit can be eliminated as the source of trouble. Other possible noise sources are the stabilized voltage supplies (particularly neon stabilizers), components (resistances, capacitances) and misalignment of the cam-operated commutator switches.

Many spectrophotometers use transistors as switches instead of electrical contacts, and the instructions in the manufacturer's manual relating to setting up and synchronization with the optical-beam switching system should be strictly adhered to. The oscilloscope waveforms that should be obtained at various points in the circuit are usually shown in the instrument manual. Examples of instruments using transistor switches are the Unicam SP700 and Zeiss PMQ II.

In optical null instruments the signal from the detector, obtained when the alternated sample and reference beams are of unequal intensity, is amplified and used to drive the recorder servosystem. The servomotor is coupled to the optical attenuator comb or wedge that is driven into the light path until the beams are balanced. In visible-ultraviolet instruments the optical attenuators are often specially designed (or a special photomultiplier circuit used) to give a direct readout of optical density. In ratio-recording instruments this is usually achieved by using logarithmic amplifiers or specially wound slide-wire potentiometers in the recorder.

Densitometers are optical null instruments and may incorporate a recording table that is driven at a speed proportional to the rate of change of density. When a dense portion of the film being scanned passes through the optical system the recording table is slowed down.

Routine maintenance on optical null instruments consists mainly of lubrication of moving mechanical parts such as film or plate table slides and rollers, cam and optical wedge carriage mechanisms, the beam switching or shutter motor, cleaning the optical components and lamp replacement. Optical null recording spectrophotometers may incorporate many mechanical parts requiring lubrication as can be seen from Figure 4.15, which illustrates some of the points requiring lubrication on the Unicam SP8000 spectrophotometer.

Faults on optical null instruments such as the pen failing to respond and recorder traces not being reproducible may be due to mechanical causes and this possibility should be checked before searching for electronic faults. For example on the Joyce Loebl microdensitometer a broken nylon drive cord may be the cause of the pen failing to move as the optical wedge cannot then balance the light beams.

Recording of wavelength scans is carried out either by fixing the paper and moving the pen across the paper by a drive connected to the wavelength scanning mechanism or by fixing the pen position and moving the recording paper under the pen on a carriage or table. The mechanics of these systems involve clutches and cable drives which may require attention if steps are seen on the recording trace. Another fault is the pen or recorder table not being firmly clamped to the drive wire so that the wavelength calibration is not reproducible. The sample signal pen chain-drive of the Perkin Elmer 257 spectrophotometer is illustrated in Figure 4.16.

Some faults on the Unicam SP8000 spectrophotometers due to the null method of their operation, are described below. The instrument is of course also subject to the faults experienced on all types of spectrophotometers such as noise from the detector, or operating circuit, and faulty lamps or alignment. Faulty pen response may be due to changes in the electronic circuit necessitating adjustment of the gain and damping controls in the amplifier, but can also be due to mechanical faults. It is important that the mechanical linkages of the servo-pen and optical attenuator system can move freely. If the zero absorbance line is not flat over the wavelength range, the internal screw adjustments on the beam balance cam require setting (Figure 4.17). Check also that the pulleys and wire-drive linking the beam balance cam, the variable attenuator comb, and the manual zero-setting attenuator comb are free and do not foul any mechanical parts. If they do then the zero line will not be reproducible. Inaccurate absorbance readings can be due to lamp or optical alignment faults but may also be due to the attenuator combs being damaged.

No response on the pen can be due to an electronic fault in the amplifier or to a broken drive cable. The same symptoms result from failure of the EHT interlock microswitch on the cell compartment cover to operate.

General checking of performance

It will be apparent from the details above that the technician, engineer or research worker who wishes to carry out adjustments or repairs must have a knowledge of both electronic and mechanical techniques. Much damage can be done by untrained inexperienced persons attempting such work. Advantage should be

SERVICING ELECTRONIC LABORATORY EQUIPMENT

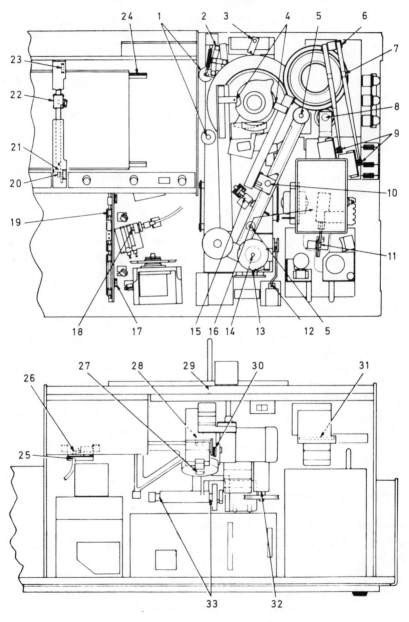

ELECTRO-OPTICAL MEASURING INSTRUMENTS

Figure 4.15. Some lubrication points on a Unicam SP 8000 spectrophotometer; above, application points; below, lubricant details. Lubrication of bearings, bearing surfaces, pulleys, wheel races, etc., should be carried out using only the lubricants listed below. The intervals suggested should be adhered to where possible.

(a) Light grease (Royal molytone 320) designated by G in table below;
(b) Light oil (SAE20) designated by 0 in table below;
(c) Antiscuffing paste containing molybdenum disulphide designated by M in table below.

The intervals at which lubrication should be carried out are indicated by the figure 6 (for 6 monthly intervals) or 12 (for annual intervals). Thus G.12 signifies that Light Grease should be applied every 12 months.

1. Chart table drive-wire pulleys (2)   G.12
2. Cam clamp mechanism pivots   G.6
3. Table link mechanism pivots   G.6
4. Prism and slit arm roller follower races (2)   0.6
5. Variable ratio pulley bearings (2)   0.12
6. Beam balance arm pivots   G.6
7. Beam balance arm cam follower rollers (2)   G.6
8. Chart shift brake plunger   G.6
9. Beam balance arm pinion engagement (2)   M.12
10. Variable ratio trolley races (3)   G.12
11. Lamp change linkage   0.6
12. OX7 filter linkage   0.6
13. Variable ratio selector disc edge   G.12
14. Variable ratio control knob spindle bearings (2)   0.12
15. Slit override screw and gears   M.12
16. Slit override control bearing and bevel gears   M.12
17. Attenuator drive-wire pulley bearings (4)   G.12
18. Set zero control screw and pivots   M.12
19. Attenuator carriage wheel races (6)   G.12
20. Pen lift catch   0.6
21. Pen carriage rail end bearings (2)   0.6
22. Pen carriage wheel races (5)   G.12
23. Pen drive-wire pulleys (2)   G.12
24. Chart table guide rails   0.6
25. Set zero control bearings (2)   G.12
26. Set zero control bevel gears   M.12
27. Servo clutch shaft bearings (2)   G.12
28. Servo worm and wheel   M.6
29. Wavelength selector plunger   0.6
30. Pen drive-wire pulley   G.12
31. Plug-in motor gears   G.12
32. Chart table drive-wire pulley   G.12
33. 3 : 1 drive pulley bearings (2)   G.12

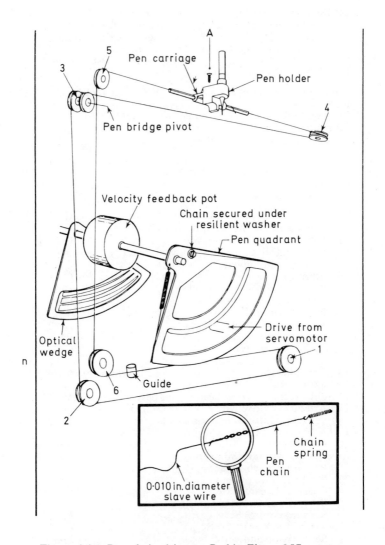

Figure 4.16. Pen chain drive on Perkin Elmer 257 spectrophotometer (courtesy of Perkin Elmer Ltd.)

ELECTRO-OPTICAL MEASURING INSTRUMENTS

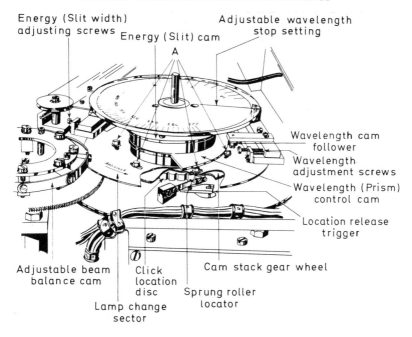

Figure 4.17. Wavelength and beam balance adjustments on Unicam SP 800 spectrophotometer (courtesy of Pye Unicam Ltd.).

taken of any training courses offered by the manufacturers and it is most important to read the instruction manual thoroughly.

When faults such as inaccurate absorbance or transmission readings, wavelength errors or excess stray light are reported it is important to check the instrument calibration and performance before attempting to trace a fault. The use of lines in a mercury or u.v. lamp spectrum for wavelength checks has already been referred to. Some useful mercury emission lines are at 253.7, 546.1, 577 and 1014 nm. With more sensitive instruments, holmium and didymium filters are also used to check the wavelength accuracy. The filter characteristics are listed in Table 4.3. Many monochromator castings incorporate thermostatically controlled heating elements so that a uniform controlled

Table 4.3. Peaks in spectrum of holmium and didymium test filters

	Wavelength mμ	Tolerance mμ
HOLMIUM FILTER	190.2	±0.4
	192.3	±0.4
	241.5	±0.4
	279.4	±0.4
	287.5	±0.4
	333.7	±0.55
	360.9	±0.75
	418.4	±1.1
	453.2	±1.4
	536.7	±2.3
	637.5	±3.8
DIDYMIUM FILTER	573	±2.3
	586	±2.3
	685	±4.5
	741	±5.5
	809	±6.3

temperature is obtained. Failure of this circuit may well cause wavelength errors. It is important to check that the thermostatically controlled heaters have been in operation for several hours before the wavelength accuracy test is carried out. Absorbance accuracy can be checked by using a standard solution checked on another instrument, a known neutral density filter or a potassium dichromate solution. Each neutral density filter has a virtually constant optical density over a range of wavelengths while the potassium dichromate solution has a precise density at certain wavelengths (e.g. 1.495 ± 0.02 O.D. at 235 nm and 0.586 ± 0.02 O.D. at 313 nm).

Holmium filters should not be used for transmission (or absorbance) calibration checks since this filter characteristic changes with temperature and the slit width used. The wavelength characteristics however are stable.

Resolution can be checked by measuring the spectrum of benzene vapour between 220 and 330 nm. There should be clear resolution of the two peaks following the large peak at 258.9 nm.

Other methods involve resolving the holmium filter doublet peaks at 382.5 and 387 nm with the valley at 384.4 nm, or using the mercury lamp doublets at 577 and 579 nm.

Linearity is checked by using a series of neutral density filters or standardizing solutions.

Stray light is checked by inserting a Corning Vycor filter (or M/20-KOH solution in a 1 cm cell) in the cell compartment and checking the light transmitted. On the SP500 this should be less than 0.2 per cent at 200 nm and on the SP8000 the absorbance indicated greater than 2 at 200 nm and 1.5 at 191 nm.

Flame emission and atomic absorption

In a flame photometer, the sample is introduced into a flame as an atomized spray and the emission of light is then dependent on the nature of the sample. A simple laboratory set-up utilizes an optical filter to isolate the narrow band of light from the sample. Photocells or photomultipliers are used as detectors. In practice, difficulties occur due to variation in the 'background' from the flame, and to the fact that the intensity of emission depends on the flame temperature.

Atomic absorption methods are usually more sensitive. In this technique, a hollow cathode lamp, its cathode containing the element to be determined, emits light with a line spectrum characteristic of the element. In passing through the flame, some light is absorbed by atoms of the sample in the flame. Thus by comparing the intensity of the light before and after the sample is introduced, a measure of the light absorption is obtained. This in turn is related to the concentration of that particular element in the sample.

As an example of an instrument capable of both emission and atomic absorption measurements the Unicam SP90 will be described. For emission work, a shutter blocks off the hollow cathode lamp, and light from the flame is modulated by a vibrating vane and passed through a monochromator to the photomultiplier detector. A diagram of the monochromator is shown in Figure 4.18. The signal is amplified and passed to a synchronous rectifier which rectifies the signal in synchronism with the light modulation. The resulting d.c. signal is displayed on

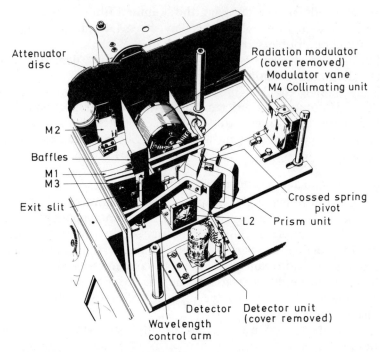

Figure 4.18. Monochromator of Unicam SP 90 spectrophotometer (courtesy of Pye Unicam Ltd.).

a meter and can be fed to a recorder. In the atomic absorption mode the hollow cathode lamp is supplied with a mains frequency interrupted d.c. supply. The detector circuit rectifier, working in synchronism with the light modulation, thus responds only to the modulated signal and ignores the continuous-flame background signal.

Several types of flame burner units are available. Common examples are nitrous oxide/acetylene, air/acetylene and air/propane. It is most important that the correct burner head is fitted for the gas being used. Failure to do this may result in a blow back in the system. Care should be exercised when using a nitrous oxide/acetylene mixture as there is a risk of flashback. On burners using air, always turn on the air before the gas, and light the burner after allowing about 10 s for the gasses to flush through the system. If this sequence is not followed a highly luminous flame results and the burner is contaminated with soot.

Should the spectrophotometer not operate, check the mains indicator lamps, fuses, and circuit voltages. No output on the meter on emission mode may be due to obstruction of the light path by a shutter. Always check this before looking for electronic faults. Noise or random fluctuations superimposed on the main meter reading may be due to electronic faults (as described earlier) a dirty or misaligned burner, a faulty atomizer or faulty modulator. Instability on the meter can be due to leaks and blockage in the gas flow system resulting in an unstable flame. Another possible cause is unstable amplitude of vibrations of the modulator.

When fault finding in the atomic absorption mode, extinguish the flame and if the fault still occurs then change the hollow cathode lamp. If this does not cure the fault check the lamp supply circuit (measure lamp current) and that the light path is not obstructed by a shutter or modulator vane (normally held clear on absorption). Further work should only be carried out by an electronic technician or service engineer. If the fault was eliminated on extinguishing the flame, probable causes are a faulty atomizer or dirty burner. A high noise level is often due to the atomizer, while low sensitivity may be due to a dirty burner or the light path not being symmetrical over the burner slots. If the hollow cathode lamp current cannot be set to the correct value, check the wiring and two matched resistors in the lamp unit. A high background on either side of the lamp emission line can be cured by activation of the lamp getter. This is done by reversing the polarity of the lamp connections and passing a current of 10 mA for 10 min. The object of this is to clean up impurities within the lamp. The connections are now reversed back to normal and the lamp run for a few hours. Lead, tin and lithium lamps are run at normal current while other lamps are run at a higher current than normal (about 28 per cent greater).

Electronic faults affect both emission and absorption modes and if no output is obtained, after checking mechanical points mentioned earlier, check the monitoring points with an oscilloscope. Start with the signal output and then check the HT supply. Next test the EHT supply to the photomultiplier with a suitable testmeter. If all the voltages are correct check through the amplifier. If low sensitivity is reported, after checking the burner and atomizer, examine the optical light path as the light may not be falling on the photomultiplier. If all these points are in order then the electronic technician should check circuit voltages and the amplifier components.

5 Recording and Indicating Devices

In this chapter the servicing of moving-coil meters, galvanometers, moving-coil recorders and potentiometric recorders is described.

INDICATING INSTRUMENTS

Moving-coil meters

In this type of meter, a coil to which a pointer is attached, is suspended between the poles of a permanent magnet and deflected by an amount proportional to the current flowing through it.

In use, meters may be subjected to electrical overload or mechanical shock. An excess of current flowing through the meter coil will deflect the pointer hard against the mechanical stop and can bend the end of the pointer so that accurate readings are impossible. In some cases it is possible to remove the meter cover and to use tweezers to straighten the pointer. Great care is, however, necessary.

Other faults which are usually the result of mechanical damage are the pointer sticking or the meter movement shaken out of its bearings. A meter movement may stick at various points on the scale, and if this is due to a small dirt or dust particle, blowing out with an air syringe may cure the trouble. The repair of meter damage is best carried out by an instrument repairer although often it is cheaper to buy a replacement meter.

Multirange testmeters such as the familiar Avo are robust instruments and incorporate an overload device that protects the meter movement and pointer from damage if excess current is passed. Instruments of this type should only be repaired by the manufacturer or an authorized meter repairer.

SERVICING ELECTRONIC LABORATORY EQUIPMENT

Meters in electronic equipment that measure voltage and current may fail to indicate correctly due to faults in the external resistors. Resistances are placed in parallel (shunt) with a meter for current measurements and in series for voltage measurements. Failure of a voltmeter to read may be due to an open-circuit resistance, and inaccurate readings to changes in the value of the series resistance. If the shunt resistance circuit is not complete on an ammeter, the full circuit current will flow through the meter movement which may be irreparably damaged. Erratic or intermittent operation of switched meter ranges is usually due to bad electrical contacts on the switch.

Moving-coil meter movements only operate with a direct voltage applied, and if an alternating voltage or current is to be measured, then a rectifier must be incorporated in the measuring circuit. Incorrect readings with oscillation of the meter pointer is a sign that the rectifier is faulty. Moving-iron meters can be used directly on a.c. circuits.

Galvanometers

Galvanometers utilize very sensitive moving-coil movements and they must be treated with care. The construction of a suspended-coil galvanometer is shown in Figure 5.1 and the coiled wires which feed the current to the coil can be seen on the left and right of the illustration. They are soldered to the main solid connecting leads. The wire-suspended moving coil, wound around the soft iron core, is in the centre of the photograph.

Galvanometers used for laboratory measurements are often of the mirror type where a small mirror attached to the movement is deflected when current flows. Light from a low voltage lamp is reflected by the mirror on to a screen so that a large light-spot deflection is obtained for a small coil-movement. The galvanometer is very sensitive and is often clamped when the instrument is moved. It is most important to release the clamp before attempting to use the galvanometer. If no light spot is visible the usual cause is failure of the lamp. When a lamp is replaced, the plate on which the lampholder is fixed may have to be repositioned in order to obtain a circular light spot in the screen.

Figure 5.1. Construction of a sensitive galvanometer.

Another reason for absence of the spot from the screen is misadjustment of the galvanometer's zero adjustor.

Failure to have the correct value of damping resistance in the external circuit of a galvanometer will cause oscillation of its pointer or light spot. The value of the resistance for a particular galvanometer is stated in the manufacturer's data and must be adhered to if oscillation is to be avoided when current is suddenly passed through the coil. If the resistance of the circuit on which measurements are being made is different from the required damping resistance, then additional resistances in a series or parallel arrangement must be incorporated. Galvanometers which have a high natural frequency do not use damping resistances but are oil damped. Changes of temperature affect the damping of this type because the viscosity of the oil changes with temperature.

RECORDERS

Moving-coil (galvanometer) recorders

There are two groups of recorder using moving coils, (1) direct writing and (2) indirect writing. In direct-writing recorders a pen or stylus is attached to the moving coil. Ink pens are often used and the principal difficulty in use is pen blockage. Dried ink on the end of the fine pen and fibres from chart paper may be the cause of the pen not following a fast input signal. In order to overcome the above problems, a heated stylus and special heat-sensitive chart paper are sometimes used. A small heated filament is contained in the stylus and failure to write is usually due to a fault in the filament circuit. On some instruments the heat applied to the filament has to be adjusted according to the chart-paper speed. If too much heat is applied on a slow speed then a wide trace appears instead of a fine thin line.

The chopper-bar recorder is another type of direct-writing instrument in which, at regular intervals, a striking bar presses the pointer against an ink ribbon positioned above the paper. This type of recorder is typically used with chromatograph column monitors. If irregularities occur in the dotted trace, check for freedom of movement by blowing gently on the pointer which should move without sticking over the full scale width. Steps in the trace may be due to fibres from the printing ribbon obstructing the pointer. If no deflection at all is obtained check the continuity of the coil and connections. On some instruments the printing ribbon is of two colours. When used with a column monitor and fraction collector in analytical chemistry, contacts in the collector operate a solenoid in the recorder every time a tube

RECORDING AND INDICATING DEVICES

is changed, so that the ribbon colours are alternated. A loud humming noise usually indicates that the solenoid is overloaded due to a mechanical fault. Failure of the ribbon to change colours may be due to the electrical contacts in the collector, an open circuit in the inter-unit connecting cable, an open-circuit solenoid or to failure of the circuit supply voltage.

Indirect-writing recorders can be divided into photographic and ultraviolet types. In the photographic type a light spot from a mirror galvanometer unit sweeps across photographic sensitive paper. The ultraviolet recorder is more complex electronically as the trace, grid lines and timing lines are produced on the sensitive paper while recording. An ultraviolet lamp and mirror galvanometer are used to produce the signal trace, while at the same time light from the lamp passes through a grid graticule to produce lines on the paper.

Timing lines are produced across the paper by a photo-flash tube which is triggered at regular intervals by an electronic circuit. This circuit comprises an accurate crystal-controlled oscillator and switched frequency dividing circuits to control the rate of triggering pulses applied to the lamp via an amplifier. If timing lines are not produced, first check the photo-flash tube by substitution. If this does not cure the fault then the electronic circuit should be checked. This is carried out by systematically checking the waveforms obtained from the oscillator, dividing circuits and trigger amplifier. If no signal trace or grid lines are produced on the paper, check the ultraviolet lamp circuit. In all probability the lamp will require replacement. If grid lines are produced but no signal trace, then the mirror galvanometer may be faulty.

Galvanometer recorders are widely used in physiological and electro-medical applications and the galvanometer coil is then driven by an amplifier coupled to a transducer. Instruments are usually multichannel and the galvanometers are mounted in a standard magnet block. Inaccuracy on several channels may be due to galvanometers having been interchanged after calibration. Since the field across the magnet block may not be absolutely constant it is important to use a galvanometer in the position in which it was calibrated.

Potentiometric recorder principle

The principle of operation of potentiometric recorders is shown in Figure 5.2. The amplifier compares the slidewire voltage with the d.c. input signal 'e', the difference being referred to as the error signal voltage. The d.c. error signal is converted to a.c. by the transformer and converter unit, amplified and used to operate the servobalancing motor. The servometer then rotates and drives the slidewire wiper, which is coupled to the recorder pen, in such a direction that the error signal is reduced to zero. The recorder is set up before use by short-circuiting the input and adjusting the zero control until the pen coincides with the scale zero. Then an e.m.f. of known value is applied, equal to the sensitivity or range of the recorder and the span control adjusted until the pen is on the scale 100 per cent position.

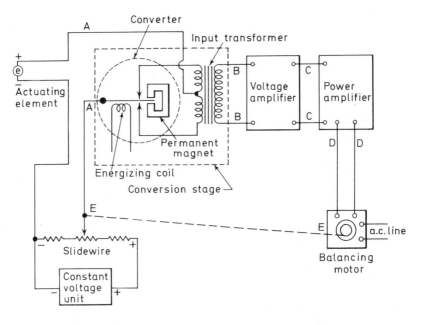

Figure 5.2. Principle of potentiometric recorder (courtesy of Honeywell Ltd.).

Maintenance of potentiometric recorders

One routine maintenance task is blowing and cleaning off dirt, particularly from the slidewire if it is of the exposed type. It is generally recommended that slidewires should be cleaned with a bristle brush, a dry lint-free cloth or paper tissues, although if very dirty, naptha or benzene can be used. Oiling exposed slidewires with contact lubricant is not always good practice as it tends to cause jitter and erratic behaviour due to dirt and dust adhering to the oil. If the slidewire is protected from the external environment, contact lubricant can be used and is recommended by some manufacturers, e.g. on Beckman laboratory recorders. The use of oil on Honeywell recorder exposed slidewires is not recommended.

The latest recorders use totally enclosed potentiometer slidewires and an example is the Honeywell Elektronik 19 recorder which is shown with the cover removed on Figure 5.3. The

Figure 5.3. Top view of Honeywell Electronik 19 Recorder (courtesy of Honeywell Ltd.).

servomotor and potentiometer can be seen in the left centre of the photograph. Cleaning and oiling of gearing, driving pulleys, etc., replacement of worn parts such as nylon drive cords, and tightening up of mechanical parts that may have loosened with use are routine maintenance tasks that should not be neglected. The chart drive gearing and rubber drive belt used on the Honeywell Electronik 19 are shown in Figure 5.4.

Figure 5.4. Chart drive mechanism of Honeywell Electronik 19 Recorder (courtesy of Honeywell Ltd.).

Ink pens are often troublesome on recorders and they should be regularly cleaned to remove dust and fibres of paper that may accumulate and cause blockage of the pen tip. If a recorder is only used periodically then, after use, the pen and capillary tubing should be washed out with warm water. If the ink is difficult to remove a weak solution of detergent can be used followed by rinsing out with water. If the ink is not cleaned out, then the pen tip will become clogged with dried ink during the period the recorder stands idle. Blocked pens can be cleaned using fine wire though care is required to ensure that the wire does not break off and become jammed in the pen tip.

RECORDING AND INDICATING DEVICES

Spilt ink should be wiped up immediately, as if left it may be difficult to remove from the recorder surfaces. After soaking up as much ink as possible, the remainder should be cleaned off with a cloth soaked in a detergent solution. Solvents should not be used as their use may damage paintwork and instrument scales.

The voltage applied to the measuring circuit is often obtained from a mercury cell which should be replaced at the intervals suggested in the recorder instruction manual. This period is typically 4 to 6 months. Mercury cells may also be used for zero off-set circuits (e.g. Telsec recorders). Stabilized zener-diode constant-voltage units are also used for the measuring-circuit voltage supply. In order to test constant-voltage units, precision resistances are connected across the d.c. output and the voltage measured as shown in Figure 5.5. This illustration shows test circuits for two different Honeywell units.

On some measuring circuits, overvoltages applied will induce a retained inductive field in the choke coil of an internal filter circuit. A special degaussing procedure is then necessary using a variable a.c. voltage source.

Noise on the recorded trace is a common fault and may be due to the amplifier gain control (usually internal) being set too high, noise from the mains supply, dirty slidewires, microphonic valves (internal electron flow from cathode to anode modulated by mechanical vibrations), defective converter or may originate in the circuit being measured. Converters of the electro-mechanical type are plug-in units and faulty operation can be verified by substituting another unit in the recorder. With photo-electric choppers, the lamp should be replaced if its operation seems irregular.

It is important for proper operation of the recorder that the gain and damping controls are correctly set. Generally the gain should be set so that, with a mid-scale deflection, manual displacement of the pen upscale and downscale always results in the pen returning to the same point. Normally this is just below the point of oscillation.

The damping control should be set so that with a step input (at least 10 per cent of full-scale voltage) applied, the pen moves without excessively overshooting the balance point. The pen

SERVICING ELECTRONIC LABORATORY EQUIPMENT

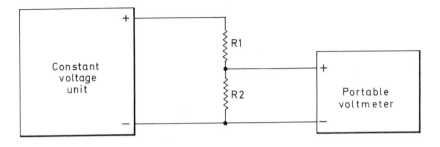

Frequency (Hz)	Current (mA)	Voltage V	Load (Ω)	Dummy load (Ω)		Voltage across R2 (mV)
				R1	R2	
50/60	6	1.029	171.5	161.5	10	60

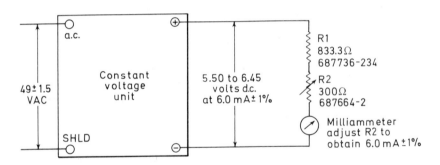

Figure 5.5. Checking constant-voltage units (courtesy of Honeywell Ltd.).

moves very sluggishly if the recorder is overdamped. This also occurs if the resistance of the circuit being measured is too high for the recorder being used.

Should the recorder have excess dead band (i.e. a large error signal must exist before the servosystem operates) the cause may be the amplifier gain too low, an amplifier fault, a defective converter or a large common-mode ripple (at both input terminals).

Failure of the pen to deflect with an input signal applied may be due to many electronic faults. A suggested sequence of fault finding is as follows: (1) check fuses; (2) check for mechanical faults if servomotor runs but pen does not move; (3) check measuring and amplifier circuit.

To check the operation of the electronic circuits, first apply a test voltage to the recorder input. If the recorder deflects then the fault is in the circuit being measured. If the recorder does not function, then check the amplifier-circuit voltages and waveforms in accordance with the information on the manufacturer's circuit diagram. If the amplifier itself appears satisfactory then the fault lies in the servomotor circuit. The servomotor should rotate in one direction if the current in its control winding (supplied from the amplifier) leads in phase the current in its reference winding (supplied from the mains) and in the opposite direction if the control winding current lags on the reference winding current. Open-circuit motor windings can be found with an ohmmeter or testmeter after switching off the supply. A fault-finding guide to potentiometric recorders is given in Table 5.1.

Chart motors

The motor driving the recorder chart to provide a time axis, usually rotates at a constant speed and different chart speeds are obtained by changing the ratio of the gearing between the motor and chart. On older type recorders this had to be done by physically changing gears, but on modern instruments this is achieved by pressing down a button connected to the gearbox by a mechanical linkage (e.g. Rikadenki recorders).

Electronic control of chart speeds is also becoming increasingly used and the chart is then driven by a stepping motor, the speed being varied by changing the frequency of the pulses applied to the motor. If the motor does not rotate, check the input pulses to the stepping motor windings with an oscilloscope. If no pulses are present then check back through the circuit of the motor drive amplifier and oscillator.

Should the motor rotate while the chart remains stationary or moves erratically, the chart drive gear wheels or motor clutch may be slipping.

Table 5.1 (a). Potentiometric recorder fault-finding guide

```
                                    ┌─ chart drive inoperative:
                                    │  Pen responds
                                    │       │
                                    │       ▼
                                    │  check voltage on chart motor
                                    │       │
                                    │       ├── bad ──► check drive switch
                                    │       └── good ─► check gearing
                                    │                      ├── good ─► replace chart motor
                                    │                      └── bad ──► repair and tighten gear drive
                                    │
              NOT WORKING ──────────┼─ pen does not respond:
                                    │  Chart motor operates
                                    │       │
                                    │       ▼
                                    │  check input signal present
                                    │       ├── bad ──► check for open circuit leads
                                    │       └── good ─► check gain and damping adjustments
                                    │                      ├── bad ──► readjust
                                    │                      └── good ─► check pen servomotor terminal voltages
                                    │                                     ├── good ─► replace motor
                                    │                                     └── bad ──► check amplifier
                                    │
                                    └─ recorder dead
                                            │
                                            ▼
                                       check fuses
                                            ├── bad ──► locate faults and repair
                                            └── good ─► check mains cable for open circuit
```

RECORDING AND INDICATING DEVICES

Table 5.1 (b)

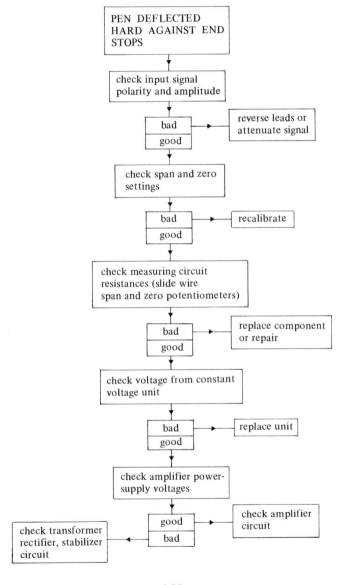

SERVICING ELECTRONIC LABORATORY EQUIPMENT

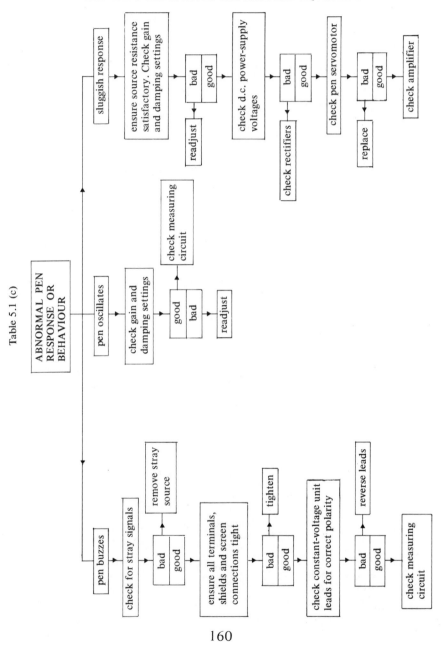

Table 5.1 (c)

XY recorders

Special types of recorder are available having two amplifier and slidewire systems so that input signals in both the X and Y axes can be plotted simultaneously. In fluorimetry for example, the output from the photomultiplier circuit measuring the intensity of the fluorescent light is applied to one axis, while a voltage derived from a potentiometer driven by the optical wavelength scanning system is applied to the other axis.

The amplifiers used are generally identical and operate on the chopper or synchronous rectifier principle. Difficulties may sometimes be experienced with nylon-cord pen-drive systems and replacement of a worn cord is often a tedious operation. It is important that the manufacturers instructions on the sequence and routing of the cord over pulleys, etc., are followed.

On some recorders a time axis is obtained by applying an internally generated linearly rising ramp voltage to the X axis amplifier.

The time-dependent voltage is usually derived from a Miller integrator circuit of the type shown in Figure 5.6. At the start of the time base sweep a negative voltage is applied to the integrator circuit input and both transistors conduct. The capacitor C1 then starts to discharge and the change of potential at TR1 collector is transferred via the capacitor to the base of TR2. This feedback action results in the rate of discharge of C1 being slowed down so

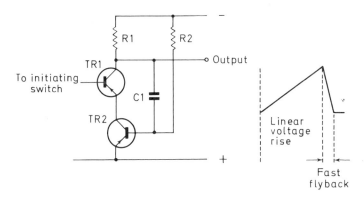

Figure 5.6. Miller time base circuit.

that a slow linear change of voltage at the circuit output is obtained. At the end of the sweep, a positive input signal cuts off TR1 and the capacitor rapidly charges through resistance R1 and the base-emitter junction of TR2, giving a rapid change of voltage at the output during the 'flyback' period.

Switches can be used to control the start and finish of the required time sweep. Different ranges are provided by switching into circuit-selected resistances.

OSCILLOSCOPES

Oscilloscope principle

The basic outline of a simple oscilloscope of the type used in school physics laboratories, or for service work where only wave shapes and not measurement are required, is shown in Figure 5.7. Electrons are accelerated away from the cathode of the cathode ray tube towards the anode and pass through the central hole in this electrode and then between deflector plates to the fluorescent

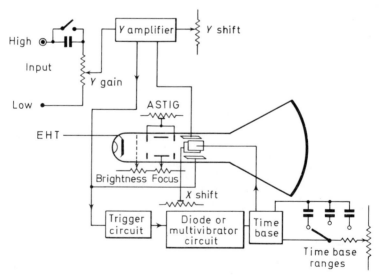

Figure 5.7. Operation of a simple oscilloscope

screen. A voltage proportional to the amplitude of the Y amplifier input signal is applied to the vertical deflection plates while the horizontal plates have a sawtooth voltage applied from a time base generator. This voltage, linear with time is usually obtained from a Miller circuit. The slope of the sawtooth waveform is varied by switched capacitors and a variable potentiometer connected in the time-base generator circuit.

When an input signal is applied to the Y amplifier, an output is taken to a trigger circuit which then produces a train of pulses. A trigger pulse is then passed through a diode to start the time base running. As soon as this happens, any further trigger pulses are prevented from reaching the time-base generator as the diode is reverse biased. When the time-base voltage has fallen to its bottom potential a second diode conducts and sets the circuit to recycle again. Some oscilloscopes use an electronic multivibrator switching circuit in place of diodes to control the Miller sweep circuit. In this type of circuit two valves (or transistors) are interconnected so that normally one stage is cut off and the other stage conducting. The trigger pulse, derived from circuits following the Y amplifier, is applied to the multivibrator and the first stage then conducts and the second stage is cut off. This initiates the time-base sweep and when the sweep has run down, conditions revert to their original state.

Many educational oscilloscopes are valve operated and, since the circuit is basically very simple, are extremely reliable. Valves of course will need eventual replacement and faults can easily be traced in conjunction with the circuit diagram. No trace or spot on the screen may be due to a cathode-ray-tube failure (check heaters) or to a power-supply fault. It should be remembered that high voltages are employed on cathode-ray tubes and in simpler instruments the negative EHT voltage is applied to the cathode from a voltage-doubling rectifier circuit. If only a horizontal line is obtained on the screen when an alternating signal is applied to the input the fault is probably in the Y amplifier. This amplifier may consist of several amplifying stages and a cathode follower output. The fault may however, be in the Y-gain control input attenuator or the input coupling capacitor (open circuit).

If the time base is not running and there is no Y input, only a

SERVICING ELECTRONIC LABORATORY EQUIPMENT

dot will be present on the screen. A vertical line is produced if an input signal is applied and there is no time base. If these faults occur on only one position of the time-base range switch, then either the switch or Miller circuit resistance or capacitance are faulty.

No time base appearing on any range may arise from a faulty trigger circuit or time base, although it should not be forgotten that the trigger circuit will not produce pulse outputs unless it is triggered off by the output from the Y amplifier.

In practice, oscilloscopes for measurement purposes are more complex than the above simple description and typically incorporate plug-in vertical amplifiers and time bases having many ranges. On these oscilloscopes it is usually necessary when changing or replacing Y amplifier valves (or transistors) to select pairs having approximately the same characteristics. If this is not done then the Y-shift control may not operate symmetrically about the centre of the screen.

Oscilloscope calibration

Due to ageing of components, oscilloscopes require recalibration at periodic intervals. This can be done by the user if he has the necessary test equipment and follows the instructions in the instrument manual. Many science and research laboratories however may not possess suitable or sufficiently accurate test equipment and in such cases the oscilloscope should be recalibrated and set up by the manufacturer. The test equipment requirements for recalibration of the Telequipment D51 oscilloscope for example, are a square-wave generator giving a frequency of 1 kHz variable between 0.5 and 100 V, square-wave generator with accurate 1 kHz 50 and 500 mV outputs, a fast-rise-time square-wave generator (less than 20 nsec rise time) giving a 100 kHz 300 mV output, a sine- or square-wave generator giving a 500 mV, 50 Hz to 1 kHz output and a 1 μsec time marker generator. In the case of a simple educational oscilloscope only a 1 kHz sine wave may be required and this should easily be obtained in the physics laboratory.

The points requiring adjustment on a general purpose oscilloscope are the input attenuators, the sensitivity and frequency compensation of the Y amplifiers, the sensitivity of the trigger circuit, the calibration of the time base, and the C.R.T. circuit.

The input attenuator is composed of fixed resistance networks with small trimmable capacitors connected in parallel with certain resistance elements. To align the attenuator a good flat-top square wave, typically 2 to 3 kHz, is applied. On each range position of the attenuator switch, the appropriate capacitor is then adjusted until the display of the square-wave on the screen has sharp corners.

The input impedance of an oscilloscope is typically represented by a 1 MΩ resistance shunted by a 50pF capacitance. In certain applications this impedance may load the circuit being tested and in such cases a high impedance probe is used. On some instruments there are compensation capacitors in the attenuator network which require adjustment with the probe connected to the input. The test square-wave is applied through the probe and the capacitors adjusted to give a flat top on the observed waveform. There is also a trimmer capacitance within the probe itself.

In dual-trace oscilloscopes, the gain of each Y or vertical amplifier is set to be the same. An additional overall-gain-balancing preset potentiometer may sometimes be fitted and allows the two trace amplitudes to be balanced with a 10 V (for example) waveform applied. The procedure, should the two traces be of unequal amplitude, is to adjust the overall gain preset potentiometer until the smaller trace occupies the required distance on the screen graticule, and then to reduce the height of the longer trace to that of the smaller by means of the appropriate individual gain preset potentiometer.

The adjustment of high-frequency compensation networks within the vertical amplifiers usually requires an accurate square-wave of 250-300 kHz with an amplitude of 200-300 mV. Then either inductors or capacitors are trimmed to obtain a displayed trace having a flat top, square corners and no overshoot.

The sensitivity of the trigger circuit is often set so that the trace is locked when the height of the displayed trace is 2 mm. Instability or erratic triggering can be due to a fault within the triggering circuit or to misadjustment of the sensitivity preset potentiometer. If an attempt is made to lock the trace with a smaller height than that recommended by the manufacturer, instability results as there is too much sensitivity.

Horizontal amplifiers are adjusted to give a certain trace-line length when there is no input to the Y amplifiers.

The time base is calibrated by feeding in square waves at various frequencies (typically 1 kHz, 100 kHz, 1 MHz) and adjusting the appropriate trimmer until one cycle of the trace occupies 1 cm on the graticule. For example with the time/cm set to 1 μsec, one cycle should occupy 1 cm at 1 MHz.

Astigmatism and geometry potentiometers are adjusted to give proper operation of the cathode-ray tube and control the potentials applied to electrodes in the tube. The astigmatism control is adjusted in conjunction with the focus. The geometry potentiometer adjusts the inter-deflection plate shield voltage and is set to correct pincushion and barrel distortion. Normally it only requires adjustment if the cathode-ray tube is changed. It should be noted that adjustment of the cathode-ray tube internal preset controls necessitates recalibration since the sensitivity of the plates is affected.

6 Radiation Detectors

Geiger-Müller Counters

Nuclear radiation is detected by alpha particles, beta particles or gamma radiation producing ionization in the detecting element. A commonly used detector is the Geiger-Müller tube which contains a central wire anode insulated from a surrounding cathode. A voltage of several hundred volts is applied between the electrodes and the tube has a low-pressure gas filling usually incorporating a halogen quenching gas. Radiation enters the tube through a thin mica end-window or through thin glass on side-window tubes. Geiger-Müller tubes have electrical connections made to an attached plug, so that the tube can then be plugged into a holder. It should be noted that the holder may house megohm resistances (connected in series with the anode circuit) and a high-voltage capacitor for coupling to the input of a scaler.

The lead from the tube holder is connected to the scaler (or ratemeter) unit which provides an EHT voltage supply for the detector, and electronic counting of the pulses produced in the detector by the ionizing particles. Geiger-Müller tubes are fragile and should be handled with care as the thin glass tube or mica end-window are easily broken if handled carelessly. A broken end-window is shown in Figure 6.1. Other faults which occur in use are leakage of air into the tube and failure of the tube quenching. The latter condition results in continuous discharge so that when used with a ratemeter, the meter pointer is deflected hard against its stop. A continuous high-pitched note will be heard from the loudspeaker in the ratemeter. If this occurs, and is due to a fault within the tube, usually a faint red glow can be seen around the anode wire.

Figure 6.1. Broken end window on a Geiger-Müller tube.

Conditions approaching continuous discharge are produced if too high a voltage is applied to the tube and the indication of this is a flood of pulses being registered on the scaler or ratemeter. Typical working voltages (within the plateau region of the pulse rate–tube voltage curve) are listed in Table 6.1. If no counts at all are registered on the counting equipment the fault may be in the Geiger tube, the wiring connections, or in the electronic circuits. The first and easiest check is to substitute a new tube. If this does not cure the trouble, then switch off and check the cable connecting the tube holder to the counter or ratemeter. Remember there is usually a resistance within the holder of several megohms. If an open circuit is found, the most likely cause is a wire broken off from one of the plug or socket connections. Intermittent counting and a background-radiation count higher than normal, may also be due to a broken connection sometimes making contact.

Scaling units

As an example of scaling equipment the Panax 102ST unit which can be used both for counting and timing experiments will be

Table 6.1. Characteristics of typical Geiger-Müller tubes

Tube	Type	Typical operating voltage V	Plateau length V	Electrical connections	Wall or window thickness mg/cm²	Use	Construction
MX168	Halogen	420	100	plug in base	3.5 to 4	beta/alpha	mica end window
MX123	Halogen	650	150	anode turret	1.5 to 2.5	beta/alpha	mica end window
2B2	Organic	1,500	–	anode pin, cathode clip	1.4 to 2.6	beta/alpha	mica end window
B12H	Organic	1,100	120	plug in base	30	beta/gamma	glass side window
B6	Organic	1,100	200	plug in base	30	beta/gamma	glass side window
MX120/01	Halogen	400	100	flying leads	525	gamma	side window
MX133	Halogen	450	100	plug in base	25	gamma	thin wall glass side window

described. An outline circuit is shown in Figure 6.2. For geiger counting, the input pulse is passed through transistors Q1 and Q2 connected as emitter followers to amplifier stage Q3. This transistor only conducts when an input pulse higher than a certain level is present, and delivers a 12 volt positive-going pulse to the first dekatron counting circuit. The dekatron tube is a gas-filled

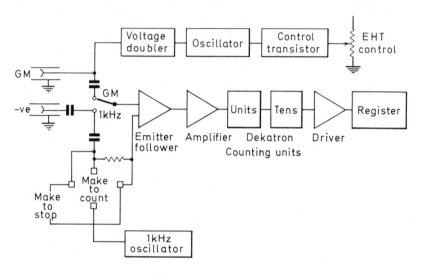

Figure 6.2. Diagram of Panax 102 ST scaler.

counting tube with one anode and ten cathodes. A discharge takes place between the anode and one of the cathodes when the tube is operating, the cathode glow being visible through the end dome of the tube. Driving pulses are applied to guide electrodes to step the glow discharge from one cathode to the next, as each 12 volt pulse is applied to the driving circuit. This circuit is a blocking oscillator and consists of a transistor and special pulse transformer which delivers negative-going 120 V narrow-width pulses to the guide electrodes. When the discharge rests on cathode nine, a transistor which is normally conducting is cut off. The transistor conducts again when the glow is stepped to the next cathode (O) so giving an output pulse to the following dekatron unit. In this way the 'tens' dekatron is stepped once for every ten pulses counted on the preceding 'units' dekatron. The output of the 'tens' dekatron stage operates a mechanical register via a transistor amplifier. The HT supply of 425 V for the dekatron anode is generated by an oscillator and voltage-doubling rectifier circuit. The Geiger-Müller tube EHT supply is obtained in a similar manner except that the output voltage level is varied by a potentiometer connected to the input of a transistor which controls the amplitude of oscillations. The remainder of the electronic circuit power supply is obtained from a conventional voltage stabilized unit. For timing experiments, the internal 1 kHz oscillator consisting of two transistors is utilized. The oscillator is connected to two sets of sockets on the front panel, one pair (make to count) when electrically connected together, cause the oscillator output to be counted on the scaler decade units (providing the count switch has been depressed). The other pair (make to stop) when connected, stop the scaler.

If, when counting a radioactive sample, no pulses are registered and it appears the fault lies in the scaler, the oscillator may be used to check the circuits. This is done by changing the function switch from the Geiger-Müller to 1 kHz position, and bridging the make to count sockets as described above. If the oscillator output is counted, then the fault is in the Gieger-Müller tube, connecting cable, or EHT supply. The tube and cable can be eliminated as described previously. This leaves the EHT supply which can be checked by connecting a high-resistance voltmeter across the Geiger-Müller socket and measuring the voltage output. If this is much lower than expected there is a fault in the voltage-doubling rectifier circuit, the transistor controlling the amplitude of the EHT oscillator or in the potentiometer control circuit. Note that most testmeters will however, slightly load the EHT supply causing a small drop in output and a 20,000 Ω/volt meter can cause the output to be 50 V low on some instruments. No high voltage at the Geiger-Müller output may be due to an open circuit in the 2M7 resistor between the socket and rectifier circuit, or to failure of the EHT oscillator.

Jumping or skipping of the glow discharge on the dekatron tubes may be due to a faulty dekatron or driving circuit. A spare plug in decade board, on which are mounted the dekatron counting tube and associated components, should then be substituted. A typical board is illustrated in Figure 6.3. If the trouble persists it may be due to incorrect supply voltages which should be measured with a testmeter. Stable voltages of +40 V–12 V and +475 V should be present. It should be noted that a fault in a preceding stage can sometimes affect a decade circuit. Failure of a dekatron to reset to zero when the reset switch is pressed, is due to switch contacts not making, a faulty dekatron, or break in the circuit resulting in the large negative potential required for resetting not being applied. Dekatron type tubes deteriorate with age and this can be seen by internal discolouration and blackening around the cathodes.

Figure 6.3. Plug-in decade counting board.

The final stage of the input amplifier of the scaler acts as a fixed-level discriminator so that only pulses greater than 0.5 V in amplitude are counted. Failure of the scaler to count on either 1 kHz or Geiger-Müller inputs may be due to a fault in this input amplifier, the power supplies or the decade units. Should the biasing of the discriminator change, then noise pulses which had previously been prevented from affecting the scaler will be passed through the circuit. This will give a higher background radiation count than normal.

More sophisticated scalers having a faster resolving time than the dekatron type (and thus able to count faster pulse rates) use transistor or integrated-circuit binary-counting decade circuits and numerical-indicator gas-discharge tubes. In numerical-indicator tubes there are an anode and ten cathode electrodes, in line and shaped to the form of the numerals 0 to 9. When a particular cathode is energized the appropriate number can be seen through the glass end of the tube.

The 'flip flop' bistable multivibrator circuit shown in Figure 6.4, is the basic element used in a counting decade. If initially

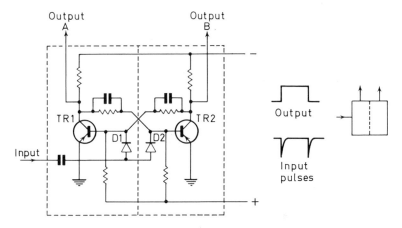

Figure 6.4. Basic multivibrator 'flip flop' counting circuit.

transistor TR1 is cut off, output A is at a negative voltage (1 state) and transistor TR2 conducts due to the circuit interconnection, so that output B is at zero voltage (0 state). This is one of the circuit's two stable states. An incoming negative pulse causes TR1 to conduct and cut off TR2, so that the circuit conditions are reversed, with a 1 state on output B and a 0 state on output A. This is the second stable state. The next input pulse causes conditions to revert to the first stable state and if we consider on output, it will seen that a square-wave output pulse is produced for every two input pulses. Thus the circuit counts in a binary scale of two mode. The incoming pulse is always applied to the cut-off stage by the pulse-steering diodes D1 and D2, since the diode connected to the conducting stage is reverse biased and therefore of high resistance. In fast scaling circuits four 'flip flops' are used in each decade unit and pulses are fed back from the last 'flip flop' to preceding 'flip flop' elements in such a way that one output pulse is obtained from the decade unit for every ten input pulses. This means that the state of the 'flip flop' elements depends on the number of input pulses applied to the decade. This is shown in Figure 6.5. Outputs from the second stage of each 'flip

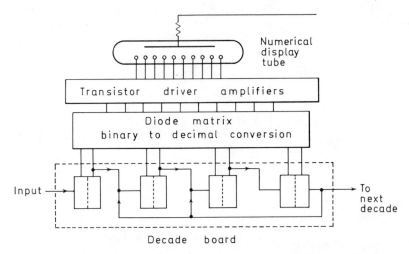

Figure 6.5. High-speed scaling decade circuit.

flop' element would then be 0 1 0 0 for a count of 4 and 1 0 0 1 for count of 9. Since the display is required in a decimal form it is necessary to use a diode matrix circuit to convert from a binary to a decimal output. Transistor driver amplifiers are connected between each number cathode of the display tube and the binary to decimal conversion circuit. Outputs may also be taken from these amplifiers to operate driving stages for the solenoids on a digital printer. There may be six decades in the scaler and if erratic behaviour or failure to count beyond a certain point is experienced on say the 10^3 decade, interchange this suspect decade with the highest count decade unit (10^6). If the fault is transferred, this will have isolated the faulty unit which can either be repaired or replaced. In the meantime the scaler can still be used on the five good decades. Each transistor or integrated circuit decade is usually a separate printed circuit board and a typical example is shown in Figure 6.6.

Figure 6.6. High-speed transistor decade board.

If interchanging the decade board does not transfer the fault, then the appropriate number display tube should be changed. Should this not cure the trouble, check the power-supply voltages to the tube and the cathode driving transistors, and test for faulty components in the relevant part of the circuit.

On complex research equipment there may be only one display unit consisting of six numerical indicating tubes, the display being switched to show the counts accumulated on one of several scalers, or the elapsed time from the timer circuit decades. In this type of equipment the isolation of a fault to a particular section can be achieved quite simply by using the selector switch. A fault present in all positions of the switch must then be in the display unit.

The ratemeter

Scaling units register the number of counts accumulated in a given time. Ratemeters however, give a visual indication of the count rate on a calibrated meter. An outline of a typical ratemeter circuit is shown in Figure 6.7. The incoming pulses are passed through an emitter follower amplifier to a pulse shaper which feeds an integrating circuit, usually of the diode pump type. Capacitor C1 charges through diode D1 during the positive part of the circuit input pulse and discharges through diode D2 during the

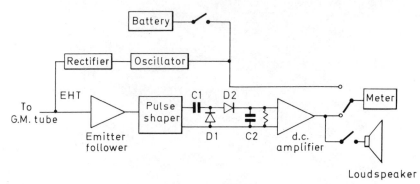

Figure 6.7. Ratemeter circuit.

negative part of the input pulse. When the rate of incoming shaped equal-charge pulses balances the leakage of charge from capacitor C2, a state of equilibrium exists and the average voltage developed across C2 is proportional to the count rate. The output from the integrating circuit is amplified and displayed on a meter calibrated in counts per min (or millirad/h for a particular isotope). When examining an inoperative ratemeter check first the state of the battery, the Geiger-Müller tube, connecting cable, and the meter movement. Ratemeters are usually portable devices and as such are liable to physical damage. If for instance the ratemeter has been dropped, then the meter movement may have been damaged. In such a case a replacement meter can be obtained from the manufacturer and fitted by the user although it may sometimes be more convenient to return the whole ratemeter to the manufacturer as recalibration may be required.

Before looking for electronic faults, if inaccurate readings or lack of sensitivity are reported, it is wise to check that the user is familiar with the instrument. Lack of sensitivity may for example be due to using the wrong type of Geiger-Müller tube for the radiation being monitored. A glass tube having a wall thickness of say 25 mg/cm^2 is not suitable for soft beta emitters but is satisfactory for ^{60}Co, ^{131}I or ^{24}Na. To monitor low-energy emitters such as ^{14}C, ^{35}S, a tube having a thin mica end-window of thickness 2.5 mg/cm^2 should be used. If it is reported that the meter indication appears lower than expected at a high count rate, check that the user understands dead or paralysis time correction. On some instruments the counting loss may be 20 per cent at full scale. Faults in the integrating circuit, stabilized voltage supplies and meter sensitivity potentiometer will give inaccurate readings.

Recalibration is necessary if for instance any of the components in the 'diode pump' circuit are changed. To recalibrate, square wave pulses are fed into the Geiger-Müller socket, through a small pico-farad capacitance. With the pulse repetition rate set to a suitable value, the preset sensitivity potentiometer in the meter circuit is then adjusted until the meter pointer indicates the correct value. It should be noted that on multirange instruments there is more than one internal sensitivity potentiometer. Typical pulse rates for calibration are 10, 100 and 1000 p.p.s. Many ratemeters that operate from the mains supply have built in test facilities, and on the test position of the selection switch the meter should indicate 50 p.p.s. Test equipment requirements for recalibration vary considerably. For instance, the Mini Instruments Model E requires negative pulses of a few volts amplitude at say 500 p.p.s., while the Nuclear Enterprises RM2 requires negative going pulses of 10 to 75 mV in amplitude to cover the range 10 p.p.s. to 10 k p.p.s.

Planchet counters

In planchet counters, beta emitting samples deposited on to paper discs are counted using a gas flow detector. The counting chamber contains an anode wire or loop, the metal body acting as the cathode. A continuous flow of gas is passed through the chamber and the radiation particles enter the detector through a thin removable window.

Incorrect operation results if the correct gas is not used or the flow rates and EHT voltages are not correct. Too high a voltage will result in continuous detector discharge and burts of counts on the scaler unit. The latter also occurs if the window material is perforated. Too high a gas pressure applied may burst the window. If the pressure and flow rate are low there is a consequent reduction in the counting efficiency. A high background count is obtained if the window or chamber is contaminated. Lack of counts may be due to an exhausted gas cylinder or leakage from connecting tubing. The scaler electronics are checked by a built-in test facility which counts the mains supply frequency (3000 counts per min).

Motor driven automatic systems employ microswitches to start and stop the sample changing mechanism and they can be troublesome in use. If jamming occurs on some systems, the mechanism clutch will slip to prevent damage to the motor and gearing but the belt drive may then be worn smooth and require replacement.

Scintillation counters

Radioactive liquids can be counted by using a liquid Geiger-Müller counter the detector tube being integral with the sample vessel which typically holds 9 ml of the sample. It is, however, important that they are thoroughly cleaned after each measurement. If they are not, then consistent and reliable results will not be obtained.

Liquid samples can be counted more effectively and with less inconvenience by using a scintillation detector. With this type of detector, a solid scintillation phosphor is excited by the radiation particles or gamma rays, and the light scintillations produced are sensed by a photomultiplier tube. The whole detector is enclosed in a light-tight and shielded assembly. The amplitude of the pulses from the photomultiplier are proportional to the energy of the incident radiation particle. These pulses are amplified and passed to a variable discriminator which can be set to reject pulses lower than a certain voltage level. The discriminator output is then passed to a scaling circuit. The photomultiplier requires a variable but stable EHT supply usually covering that range 0–2 kV. A problem encountered with this type of equipment is a high and variable background radiation count. This can be due to leakage of light into the counting chamber, electrical noise, dirty contacts on the photomultiplier plug pins and socket, bad soldered joints or screen connections on the cable from the scintillation detector assembly to the scaler/discriminator, a fault in the EHT supply, or to thermal noise from the photomultiplier. Probes incorporating a scintillator crystal and photomultiplier are available for use with some ratemeters and are useful for monitoring low-energy isotopes due to their high detection efficiency. A photomultiplier with a '2 or 3' thalium-activated sodium iodide crystal is commonly used to detect gamma radiation. This type of crystal is hygroscopic and, except where coupled to the photomultiplier, may be enclosed in a silver foil to exclude moisture which would otherwise damage the crystal. The foil also acts as a light shield.

In liquid scintillation counting, the sample dissolved in a suitable solvent and a liquid scintillator are all contained in a glass or plastic vial. In simple systems a single photomultiplier is optically coupled to the bottom of the glass vial and the resulting pulses from the photomultiplier are amplified and fed to discriminator and scaling units. In practice, when counting low-energy isotopes with such a system the background thermal noise pulses from the photomultiplier cathode are of the same order as the pulses due to the sample and it is necessary to reduce the background count by cooling the photomultiplier. Phosphorescence of samples if exposed to ultraviolet light is troublesome and it is necessary to 'dark-adapt' the samples for say 30 min before counting.

All modern automatic liquid scintillation counters now use two photomultiplier tubes which view the sides of the vial. The tube outputs are connected to a coincidence circuit which only allows a pulse to be counted if it is sensed by both photomultipliers simultaneously. Unwanted pulses due to thermal noise, etc., are of a random nature and do not result in simultaneous outputs from both photomultipliers. A high background count on equipment of this type can be due to the photomultiplier tubes or a fault in the coincidence circuit or the EHT supply but is more likely to be due to contamination of the counter chamber or the entry of light. Several discriminator circuits and scalers are usually used. The discriminators can be regarded as voltage-level controls (proportional to pulse height and therefore radiation energy) and can be set so that one scaler counts pulses between say levels A and B, another between C and D and a third between E and F. In this way it can be arranged for example that pulses due to ^{3}H are counted between levels A and B and pulses due to ^{14}C fall mostly between C and D. There is of course some overlap.

A change in counting characteristics can be due to drift in the EHT supply, to the photomultiplier, or to a fault in the discriminator/amplifier circuits. Before calling for expert electronic advice, when a user reports such a fault, the laboratory technician should ensure that it is not the user's sample that is at fault (e.g. high internal quenching) or the user himself (incorrect instrument settings).

Automatic counting systems

In research and routine analysis, large numbers of radioactive samples may have to be counted and this has led to the development of several different types of sample changing systems. In liquid scintillation and gamma counting, the vial or tube containing the sample must be transported to a position above the counting chamber, lowered into the chamber, counted, returned to the transport system, and this sequence repeated for the subsequent samples. Generally there are two types of sample-handling systems, (1) those using a conveyor belt in which the vials are placed and (2) the handling of samples in groups of say ten in a tray.

Many systems use microswitches to stop the conveyor belt or tray in the correct positions, and to operate the motorized elevator system which lowers and raises the sample from the counting chamber. These microswitches are often a source of operating difficulties, not necessarily because of electrical faults but due to failures in the mechanical actuation system. Their positioning is often critical and replacement of microswitches may require setting up adjustments and procedures which are known only to the manufacturer's service engineer. Maintenance by the user's technician, unless he has been specially trained, is thus restricted to keeping the sample changer clean, and lubrication, where required, of moving parts. Typical systems using microswitches are those manufactured by Nuclear Chicago and by Philips (Pye-Unicam).

The electronic circuits in automatic scintillation counting systems are fairly complex and include not only scaling, timing, amplifier, stabilized power supplies and read-out circuits but logic control and computing and calculating circuits. Servicing should only be carried out by an engineer or electronic technician who has been instructed on the circuit techniques used. Much damage can be done by inexperienced or unqualified persons attempting to rectify faults without knowing what they are about.

The laboratory technician and research worker should of course be familiar with the principles involved and fully understand the functioning of all the operating controls. Routine tasks such as correct fitting of new ribbons on printers, cleaning sample changers, defrosting refrigeration systems, lubricating gearing with the recommended grade of oil, replacing fuses, etc., are all within the capabilities of the laboratory technician, but more involved work as stated above should be left to a trained electronics technician.

Detailed instructions on routine maintenance tasks are given in the instrument instruction manual. In some cases methanol should be used to clean the sample conveyor, as detergents and water may affect lubricants used. If a glass vial has broken, remove the glass fragments with tweezers, and mop up the spilt radioactive liquid immediately. Decontamination of the parts affected should then be carried out. A good decontaminant is Decon 75. If a conveyor belt or chain is removed for cleaning, ensure that it is put back correctly so that the read-out of sample number agrees with the conveyor sample number in position over the elevator mechanism.

When cleaning the sample changer it is always good policy to check the shutter mechanism and light seals in the top of the elevator assembly.

Listers or printers and teletype machines being mechanical devices can be troublesome and in order to reduce wear to a minimum, dust and dirt should be blown out and moving parts lubricated regularly. A fine hair brush or damp lint-free cloth are also useful when cleaning. Felt washers and oilers should be saturated with oil while most other parts require only 2 or 3 drops. It is most important to follow the manufacturer's instructions regarding grades of oil and grease. Care must be taken to avoid getting oil on areas which must be kept dry. Care should also be exercised when cleaning, not to alter the relative positions of cams, wheels, levers, etc. The complex mechanism of a printing calculator is shown in Figure 6.8. Solvents should not be used on plastic materials as some types attack plastics.

Figure 6.8. Complex mechanism of a printing calculator.

Should a printer ribbon be fitted incorrectly or the spool mechanism fail to reverse then subsequent operations of the printer by the electronic control circuit will cause the ribbon to be torn. The fragments of ribbon within the mechanism should be removed when fitting a replacement ribbon, or they may interfere with the correct operation of the printer.

Refrigeration systems are defrosted by switching off the electrical supply to the refrigeration circuit and when the ice has softened, removing it with the aid of a blunt tool. The heat-exchanger radiator may require periodic cleaning with a bristle brush or vacuum cleaner in order to maintain an efficient air flow. Refrigeration compressor motors are hermetically sealed and require no lubrication, although the fan motor bearings do require lubrication at about six monthly intervals. If a refrigeration compressor motor is overloaded, due for instance to continuous operation in a high ambient temperature, a thermal cut-out situated on the motor operates and cuts off the power to the motor.

Some automatic sample-handling systems use a lamp and miniature photocells in order to sense and control sample vials. The points to check in such systems, if the sample changer operation is incorrect, are the lamp and photocell positions and condition. Beckman and Wallac scintillation counters are examples of such systems. The Wallac system has a pneumatic operated sample changer and it is important that the water separator is emptied once a week (more often in hot humid weather). Maintenance and cleaning of the magnetic valves and piston cylinders necessitates stripping these components and this should only be done by a trained service man. A useful feature of this system is a built-in sample-changer test panel. A switch on the panel allows the sample changer to be operated by start/stop, forward/reverse, sample and slide switches on the test panel. It can therefore be quickly established whether an operating fault is due to the electronic controlling circuits in the programmer unit or to a fault in the sample changer itself.

Appendix

BOOKS FOR FURTHER READING

ACKERMAN, P. G. *Electronic Instrumentation in the Clinical Laboratory,* Little, Brown & Co., Boston
DOBOS, D. *Electronic Electrochemical Measuring Apparatus,* Terra, Budapest
KING, G. J. *Rapid Servicing of Transistor Equipment,* G. Newnes, London
LEE, L. W. *Elementary Principles of Laboratory Instruments,* The Mosby Co. Saint Louis
MIDDLETON, R. G. *Troubleshooting with the Oscilloscope,* Howard W. Sams, New York
PRICE, L. W. *Electronic Laboratory Techniques,* J. & A. Churchill, London
PRICE, L. W. *Practical Course in Instruments and Electronics,* United Trade Press, London
RISSE, J. A. *Understanding Electronic Test Equipment,* Foulsham Sams, London
SCHUSTER, D. H. *Logical Electronic Troubleshooting,* McGraw Hill, Maidenhead, Berkshire
WINSTEAD, M. *Instrument Check Systems,* Lea & Febiger, Philadelphia
WEISSBERGER, A. and ROSSITER, B. W. *Physical Methods of Chemistry,* Wiley-Interscience, New York
ZUCKER, M. H. *Electronic Circuits for the Behavioral and Biomedical Sciences,* W. H. Freeman, San Francisco

APPENDIX

DATA BOOKS AND TABLES

Electrothermal 'vade mecum', Electrothermal Engineering, London
Handbook of Transistor Equivalents and Substitutes, Babani Press, London
Isco Tables (Handbook of data for biological and physical scientists), Instrumentation Specialities Co. Inc., Lincoln, Nebraska
Radio Valve & Transistor Data, Iliffe, London
Semicon Transistor Data Manual, Functional Publication Services, Wokingham, Berkshire
Zeus Precision Data Charts and Reference Tables for Drawing Office and Workshop, Buck & Hickman Ltd., London

ELECTRICAL AND ELECTRONIC DATA

Direct current formulae

Ohms Law: $E = I \times R$; where E = voltage (volts), I = current (amperes), R = resistance (ohms).

Power P (watts) $= E \times I = I^2 R = E^2/R$

Series Circuits: total resistance $R_T = R_1 + R_2 + R_3 + \ldots$

Parallel Circuits: total resistance $\dfrac{1}{R_T} = \dfrac{1}{R_1} + \dfrac{1}{R_2} + \dfrac{1}{R_3} + \ldots$

Time constant $t = RC$ and is the time required to charge a capacitor to 63 per cent of its final value where C = capacitance (farads) and R = resistance (ohms).

Alternating current formulae

Symbols: L = inductance (henries), C = capacitance (farads), f = frequency (hertz), $\omega = 2\pi f$, Z = impedance (ohms)

Impedance is opposition to flow of alternating current

APPENDIX

In a purely inductive circuit, Z = inductive reactance $X_L = \omega L$ and the current lags 90° in phase behind the applied voltage

In a purely capacitive circuit, Z = capacitive reactance $X_c = 1/\omega C$, and the current leads the applied voltage by 90°.

Series capacitive circuit: total capacitance

$$\frac{1}{C_T} = \frac{1}{C_1} + \frac{1}{C_2} + \frac{1}{C_3} + \ldots$$

Parallel capacitive circuit: total capacitance

$$C_T = C_1 + C_2 + C_3 + \ldots$$

If both resistance R and reactance X are present, the impedance

$$Z = \sqrt{(R^2 + X^2)}$$

With resistance, capacitive and inductive reactance,

$$Z = \sqrt{\left[R^2 + \left(\omega L - \frac{1}{\omega C}\right)^2\right]}$$

In a parallel a.c. circuit, the total current flowing is the vector sum of the individual current paths,

$$I_T = \sqrt{(I_1^2 + I_2^2)}$$

True power = $I^2 R = E \times I \times$ power factor

Power factor = cosine of phase angle ϕ between applied voltage and resultant current: $\cos \phi = R/Z$

Root mean square (r.m.s.) voltage = *maximum (peak) voltage*$/\sqrt{2}$

Transformer ratio:

$$\frac{\text{primary winding turns}}{\text{secondary winding turns}} = \frac{\text{primary voltage}}{\text{secondary voltage}}$$

APPENDIX

Current capacity of equipment wire
PVC insulation, rating typically 1 kV r.m.s. at 70°C.

size mm	spec. D.E.F61-12 rating amps	commercial rating amps
1/0.6	1.8	3
7/0.2	1.4	2
16/0.2	3	4
24/0.2	4.5	6
32/0.2	6	10

British Standard BS 1852 resistance code
Letter represents decimal point location as shown below.

R used for values up to 100 ohms, K for thousands and M for millions of ohms.

The final letter of the code indicates tolerance: F = 1 per cent, G = 2 per cent, J = 5 per cent, K = 10 per cent, M = 20 per cent.

Examples: 4.7 Ω 10 per cent is 4R7K
            0.33 Ω 20 per cent is R33M
            6.8K Ω 2 per cent is 6K8G
            4.7M Ω 20 per cent is 4M7M

ELECTROCHEMICAL DATA

Slope Factor : pH Measurements

Temperature °C	15	20	25	30
mV/pH unit	57.17	58.17	59.16	60.15

Some Standard Buffers for pH Meter Calibration

Potassium hydrogen phthalate
0.05 M. Dry crystals at 110°C. Dissolve 10.12 g in water and make up to 1 litre.

pH 4.005 at 25°C
pH 4.001 at 20°C

APPENDIX

Disodium hydrogen phosphate + potassium dihydrogen phosphate
0.025 M. Dry salts at 120°C and use 3.53 g Na_2HPO_4 + 3.39 g KH_2PO_4 in water (CO_2 free) to make 1 litre.

>pH 6.862 at 25°C
>pH 6.878 at 20°C
>pH 6.84 at 37°C

0.5 M. Use 22.4 ml of 0.5 M-KH_2PO_4 + 25.8 ml of 0.5 M-Na_2HPO_4 diluted to 1 litre

>pH 7.00 at 25°C.

Hydrochloric acid and potassium chloride, 0.2 M
Use 201 ml of 0.2 M-HCl + 299 ml of 0.2 M-KCl diluted to 1 litre.

>pH 1.5 at 25°C.

Hydrochloric acid, 0.1 M
>pH 1.10 at 25°C and 20°C.

Sodium hydroxide, 0.1 M
>pH 12.88 at 25°C
>pH 13.6 at 20°C.

Borax ($Na_2B_4O_7 . 10H_2O$), 0.1 M
>pH 9.225 at 20°C
>pH 9.180 at 25°C.

Calcium hydroxide, saturated solution
>pH 12.627 at 20°C
>pH 12.454 at 25°C.

Potassium tetraoxalate, 0.05 M
>pH 1.679 at 25°C
>pH 1.675 at 20°C.

APPENDIX

SPECTROPHOTOMETRIC DATA

Standard Solutions for Photometric Calibration

Potassium chromate, 0.04 g/litre in 0.05 M-potassium hydroxide
Values given for 25°C in 10 mm silica cell:

%T	0	35.8	17.5	70.9	27.6	10.2	40.2	92.7	100
A	–	0.45	0.76	0.15	0.56	0.99	0.4	0.3	0
nm	210	220	275	300	350	375	400	450	500

Potassium dichromate, 60 mg/litre in 0.005 M-sulphuric acid
Values given for 10 mm silica cell.

%T	17.9	13.5	50.9	22.7
A	0.75	0.87	0.29	0.64
nm	235	257	313	350

Recommended Solid Filters

(a) For transmission and absorbance calibration

Neutral density type: typical figures are given below, but manufacturers give measured values on actual filters supplied.

Kodak ND 1.0

A	1.5	1.13	1.06	1.04	1.02	1.04	1.02
nm	400	450	500	550	600	650	700

Gilford filter set. Values measured at 550 nm
Absorbances: A 0.101 ± 0.01, 0.953 ± 0.01, 1.987 ± 0.02.

(b) For Wavelength calibration

Use holmium and didymium filters; for details see text.

Useful Emission Lines from Mercury Lamp in Quartz Envelope

nm 253.6, 302.2, 334.2, 404.7, 435.8, 546.1,
    578 (576.96 + 579.07)

APPENDIX

Units and Formulae

One angström, Å = 10^{-10} metre
One nanometre, nm = 10^{-9} metre

Per cent transmittance $T = \dfrac{\text{intensity of emerging light}}{\text{intensity of incident light}}$

Absorbance, $A = \log_{10} \dfrac{1}{T}$

Absorbance–transmission conversion

A	0	0.05	0.1	0.15	0.2	0.22	0.3	0.4	0.5	0.52
$\%T$	100	90	79	70	63	60	50	40	32	30

A	0.6	0.7	0.8	0.9	1.0	2.0	3.0	4.0
$\%T$	25	20	15.8	12.6	10	1.0	0.1	0.01

CONVERSION FACTORS AND PHYSICAL DATA

1 inch = 2.54 cm
1 mm = 1000 microns (μm) = 10^7 Å
1 lb = 0.45 kg, 1 gallon water = 10 lb
1 cu. ft water = 6¼ gallons (British) = 7.48 gallons (U.S.A.)
1 horse power = 746 watts
1 litre = 0.0353 cu. ft = 0.22 gallon (British)
1 pint = 20 fluid ounces = 568 ml
1 mm Hg = 1 Torr (to within 1 part in 7×10^6)
760 Torr = 1 standard atmosphere
1 atmosphere = 14.7 lb/sq. inch = 1.034 kg/sq. cm
0°C = 32°F = 273°K

$°C = (°F - 32) \times \dfrac{5}{9}; \quad °F = \left(\dfrac{9}{5} °C\right) + 32$

Ranges of temperature measuring devices
Platinum resistance thermometer −259 to +630°C
Mercury in glass thermometer −38 to +600°C
Negative temperature coefficient thermistor −80 to +150°C

APPENDIX

Thermocouples, typical ranges
Copper constantan -200 to $+300°C$; e.m.f./°C $35\mu V$ at $20°C$, $17 \mu V$ at $-200°C$
Silver palladium $+180$ to $+630°C$; e.m.f./°C $17 \mu V$ at $180°C$, $30 \mu V$ at $630°C$
Platinum, 10 per cent rhodium platinum 0 to $+1250°C$; e.m.f./°C $7 \mu V$ at $100°C$, $13 \mu V$ at $1000°C$
Chromel alumel -100 to $+600°C$; e.m.f./°C $30 \mu V$ at $-100°C$, $40 \mu V$ at $+600°C$

Index

Aerosol freezers, 23
Aerosol switch cleaner, 24-7
Amino acid analysers, 98
Amplifiers,
   audio, 70-3
   chopper, 82
   class A and B, 70
   direct coupled, 82
   push pull, 70
   testing, 71-2
   transistor, 11
   valve, 7
Antisurge fuses, 2
Arcing, 24
Armature, 42
   testing, 44
Atomic absorption, 143-6
Attenuator, 68
   alignment, 165
Automatic sample changers, 180-88
Automatic temperature compensation, 79

Ballast valve, 56
Barrier layer cell, 116
Batteries, 38-9
Bearings, centrifuge, 54
Bearings, motor, 44
Beckman
   gas chromatograph GC2A, 106
   hydrogen/deuterium lamp supply, 114
   spectrophotometer DK2, 131-2
   ultracentrifuge L265, 50

Bellows thermostat, 30
Bench centrifuges, 47
Benzene vapour test, 142
Bimetal thermostat, 29
Binary scale of two, 173
Bistable multivibrator, 173
Bleeder chain, 67
Braking circuit, 53
Brushes, 44, 47
Buffer solutions, 75
Burners and gases, 144

Calibrating
   oscilloscopes, 164-6
   pH meters, 84
   ratemeters, 177
   recorders, 155-7
   spectrophotometers, 141-3
Capacitor,
   colour code, 4
   discharging, 5
   electrolytic, 4
   time constant, 74
Carbon dioxide electrode,
   principle, 88
   testing, 88-9
Cartridge fuses, 2
Cathode, 6, 8
Centrifuge,
   bearings, 54
   bench, 47
   brushes, 47
   heads and rotors, 54-5
   speed control, 45-9

INDEX

Centrifuge—*cont.*
  ultra and high speed, 49-54
Chart drive motors, 157
Chopper amplifier, 82
Chopper bar recorder, 150
Chromatography,
  gas, 99
  liquid, 94
  principles, 93
Circuit symbols, 15-17
Cleaning
  electrical contacts, 24-6
  mirrors, 118
  pens, 154
  recorders, 155
  sample changers, 181
Cold cathode valve, 8
Colorimeter, 99, 118-19
Colour codes,
  capacitors, 4
  resistance, 3
  wiring, 14, 18
Column,
  cleaning, 106
  monitors, 119
Common emitter amplifier, 11
Commutator maintenance, 44-5
Component damage, 5, 6
Constant current power supply, 63
Constant voltage
  power supply, 62
  transformer, 59
  unit, 155-6
Contact
  cleaning, 24-5
  lubricants, 26
  thermometer, 31-3
Contactor, 53
Converters,
  electro-mechanical, 82
  photo-resistive, 82
  vibrating capacitor, 85
Corning Vycor filter, 143
Corroded drop counter, 98
Corroded switch contacts, 126

Damage to
  capacitors, 2
  centrifuge rotor, 55
  resistors, 3
  transformers, 2
  transistors, 14
Damping, 149-50, 155
Dead band, 156
Decade counters, 170-75
Decontamination, 181
Dekatrons, 170
Densitometers, 111, 136
Deuterium lamp, 113
Didymium filter, 142
Diode,
  semiconductor, 9-10
  testing, 20
  valve, 5, 7
  zener, 9
Direct current power supply, 60
Discharging capacitors, 5
Discriminators, 172, 178
Drift, 82
Drop counter, 96-8
Dropping mercury electrode, 91
Dry batteries, 38-9
Dry joint, 23
Dust and leakage, 65-6

Electrical contacts, 24
Electrochemical electrodes, 90-91
Electrolube, 26
Electrolytic capacitors, 4
Electrometer amplifiers, 78, 82
Electrometer valves, 78, 127
Electron flow, 5
Electronic
  Instrument Vibret pH meter, 85
  thermometers, 36-8
  timers, 74
Electrophoresis, 65-7
Electrostatic effects, 86
Emitter follower, 68
Enclosed slide wire, 153
Energy regulators, 31

INDEX

Equivalent
  batteries, 39
  transistors, 12
  valves, 8
Exposed slide wire, 153
Extra HT supply, Linstead, 64

Fault finding guide,
  flame ionization gas chromatograph, 105
  null balance spectrophotometer, 123-5
  pH meter, 79-81
  potentiometric recorder, 158-60
  ratio recording spectrophotometer, 133
  thermal conductivity gas chromatograph, 109
Fault finding logic, 22
Fault, physical indications, 2, 3
Field effect transistor, 82
Filters,
  Corning Vycor, 143
  didymium, 142
  holmium, 142
  neutral density, 142
  potassium dichromate, 142
Flame emission spectrophotometer, 143
Flame ionization
  detector, 99-100
  gas chromatograph fault finding, 105, 109
Flip flop, 172
Fraction collectors, 95-8
Freezers, 23
Frequency measurement, 69
Frequency response tests, 71-2
Full wave rectifier, 61-2
Fuses, 1

Gain adjustments, 156, 165
Gain amplifier, 155
Galvanometer construction, 149
Galvanometer damping, 149

Gas chromatograph,
  Beckman GC2A, 106
  column
    cleaning, 106
    oven temperature control, 102-3
  electron capture detector, 104, 107
  flame ionization detector, 99-101
  interpretation of faults, 108-10
  Pye 104, 99-102
  thermal conductivity detector, 107-9
Geiger Müller
  counters, 167-75
  tube faults, 167-8
  tube characteristics, 169
Generator, signal, 67-8
Generator, tacho, 48
Glasses, u.v. protective, 115
Globar rod, 115
Gold electrode, 90
Grid current, 78

Heated stylus, 150
High impedance probe, 21
High intensity lamps, 115
High speed decade circuit, 174-5
High voltage electrophoresis, 66-7
High voltage power supplies, 63
Hollow cathode lamps, 145
Holmium filter, 142
Hot wire vacuum relays, 31
Hydrogen lamp, 112-13

Impedance, 73
Impulse counter, 96
Induction motors, 42-3
Infrared sources, 115
Ink pens, 154
Insulation testing, 27-8
Integrated circuits, 23
Interference, 49
Interlock circuits, 52, 117
Intermittent faults, 24

Lamps,
 alignment, 127
 deuterium, 113, 137
 housing, 130
 hydrogen, 112-13
 quarts halogen, 112
 tungsten, 111-12
 xenon, 115
Leakage current, 14
Limit switches, 87
Linearity, 143
Linstead Electronics EHT supply, 64
Linstead signal generator, 68
Liquid pumps, 94
Liquid scintillation counter, 179-80
Logical fault finding, 22
Lubrication, 154

Maintenance, 28
Majority current carriers, 9
Manganese alkaline battery, 38-9
Mechanical timers, 73
Megger, 27
Mercury battery, 38-9
Mercury lamp, 141
Metal screening, 83
Microphonic valves, 155
Microswitches, 31, 97
Miller time base, 161-2
Mini instruments Model E, 177
Minority current carriers, 9
Mirror cleaning, 118
Monochromators, 118
Motor brushes, 44
Motors, 41-8
Moving coil meters, 147
MSE super minor centrifuge, 48
MSE super speed 75 ultracentrifuge, 52
Multirange testmeters, 18
Multivibrator, 163, 173

Negative feedback, 71
Nernst glower, 115-16
Neutral density filter, 143

Nichrome infrared source, 115
Nuclear Enterprises RM2, 177
Null balance spectrophotometer, 120-28
Numerical indicator tubes, 173-4

Optic-electric instruments, 111-46
Optical null spectrophotometers, 136-7
Oscilloscope,
 calibration, 164-6
 principle, 162-4
 use, 20-21
Overvoltages, 155
Oxygen electrode, 89
Oxygen measuring apparatus, 89-90

Panax 102ST scaler, 170-71
Pen cleaning, 154
Pen drive system, 140
Pentode valve, 5-7
Perkin Elmer 257 spectrophotometer, 132
pH,
 electrodes, 76-7
 meters, 77-85
 practical definition, 75
 temperature relationship, 75
Photoconductive cell, 116
Photoelectric cells, 116-17
Photoelectric drop counters, 96-8
Photographic galvanometer recorder, 151
Photomultipliers, 116-17
Photosensitive resistor, 82
Physical indications of faults, 2-3
Pirani gauges, 57
Planchet counters, 177-8
Platinum electrodes, 90
Polarograph difficulties, 91
Positive feedback, 67
Potassium dichromate, 142
Potentiometric recorder,
 fault finding guide, 158-60
 maintenance, 153

INDEX

Potentiometric recorder—*cont.*
   principle, 152
Power control circuit, 47
Printers, maintenance, 181-2
Probe, high impedance, 21
Proportional control, 40, 87
Pulse polarograph, 91
Pumps, liquid, 94
Pumps, peristaltic, 94
Push pull amplifier, 70
Pye 104 gas chromatograph, 99-102
Pye 78 pH meter, 77-8
Pye 290 pH meter, 84

Quartz halogen lamp, 112
Quartz lamp window, 113

Ratemeter, 175-7
Ratio recording spectrophotometer, 132
Reactivating pH electrodes, 76
Rectifier,
   full wave, 61
   principle, 5
   ratings, 62
   smoothing, 61
Reed switch, 128
Refrigeration unit control, 39
Refrigeration unit motors, 183
Relay 24
Replacement printed circuit boards, 33
Repulsion motor, 42
Resistance colour code, 3
Rheostat motor control, 45
Routine maintenance, 28

Safety, 66
Scaling circuits, 168
Scintillation counters, 178-80
Screening, 89
Semiconductors, 9-14
Signal generators, 67-9
Silica gel, 121
Silicon controlled rectifier, 32
Silver electrode, 91

Silver/silver chloride electrode, 91
Simmerstats, 29
Slidewire cleaning, 153-4
Slow blow fuses, 2
Soldering transistors, 11-12
Solenoid valve, 86
Sparking, 49
Specific ion electrodes, 92
Spectrophotometers,
   general, 111-46
   null balance, 120-29
   optical null, 136-41
   ratio recording, 129-35
   testing, 141-3
Speed control, 41-7
Speed measurement, 48
Stem type thermostat, 29-30
Stepping motor, 94-5, 157
Stray light, 135
Stray signals, 35
Stroboscope, 49
Switch cleaners, 26
Symbols, 15-17
Synchronous motor, 43

Teletype machines, 181
Testing,
   carbon dioxide electrodes, 88
   components, 19
   continuity, 27
   insulation, 27
   motors, 28
   transistors, 13-14
   valves, 8
Testmeter, multirange, 18
Thermal conductivity detector, 106-9
Thermal fuse, 102
Thermocouples, 34-6
Thermometers, contact, 31
Thermometers, electronic, 36-7
Thermostat,
   bellows type, 30
   bimetal, 29-30
   stem type, 29-30

INDEX

Thyratron, 33
Thyristor, 32
Time base, 163
Timers, electronic, 74
Timers, mechanical, 73
Titration equipment, 86
Toluene, 66
   temperature regulator, 33
Tracking, 43
Transformers, constant voltage, 59
Transformers, variable, 58
Transistors,
   circuit operation, 10-11
   damage to, 14
   field effect, 82
   power, 12
   principles of, 9
   soldering of, 11-12
   substitution of, 12
   unijunction, 74
Triode valve, 5
Tungsten electrode, 91
Tungsten lamps, 111-12

Ultracentrifuges, 49-51
Ultraviolet
   lamps, 112-13
   protective glasses, 115
   recorders, 151

Unicam spectrophotometers,
   SP90, 143-6
   SP500.1, 121
   SP500.2, 128
   SP600, 120-22
   SP800, 141
   SP8000, 137-9
Unijunction transistor, 74
Universal motor, 42

Vacuum leak tracing, 58
Vacuum measurement, 57
Vacuum pumps, 56
Valves, connections to, 8
Valves, faults in, 8
Valves, principles, 5-7
Vibrating capacitor, 85

Waveform distortion, 60
Weinbridge oscillator, 68
Wheatstone bridge, 36
Wiring colour codes, 14, 18

Xenon lamps, 115
X-Y recorders, 160

Zinc carbon batteries, 38-9